Introduction to Physics

Volume 1: Mechanics

First Edition

Martin Nikolo, Ph.D.
St. Louis University

Copernicus Press

First printing 2018

ISBN 0-9741638-6-4

LCCN 2003-107284

Table of Contents

Chapter 1
Measurement and Motion

Motion is fundamental to our understanding of nature. Even when we cannot see motion it is everywhere: the constant vibration of atoms and molecules that form our bodies and the Earth's spinning as it moves around the Sun are invisible to the naked eye, as is the expansion of the universe. Motion is also essential to life on Earth: many animal species, like the cheetah, depend on it to survive, as when they use their speed to hunt for food.

In this chapter we will examine how physicists measure and describe motion by defining displacement, speed, velocity, and acceleration. This will lay a foundation that will allow us to study other topics, such as force and energy, in later chapters.

We will present visual learning approach by drawing pictorial representation of physics situations and problems, motion diagrams, and graphs. We will combine these elements to enhance the visual learning experience in this and later chapters.

1.1 Physics and Measurement

The word **physics** has its origin in the Greek word *physis,* which means nature. It is a natural science that examines matter, energy, space, and time. Ancient Greek civilization dating as far back as 600 BCE gave rise to a long line of philosophers, who developed and applied mathematics and logic to explain their observations of natural phenomena. Physics as a science did not start to emerge until much later, during the 16th and 17th centuries. It was no coincidence that the improved glassmaking and lens-grinding techniques in the late 16th century brought about the invention of a telescope. The positions of planets and stars in the heavens were recorded with improved accuracy and the new data lead to many discoveries (Chapter 3).

physics The science and study of matter, energy, and their interactions.

Today physics covers a wide range of phenomena, from the smallest sub-atomic particles, to the largest galaxies. We call it the *fundamental science*. Its principles form a foundation of all other sciences.

What Is Physics?

Physicists search for answers to questions ranging from when the universe began to why we have abrupt weather changes or why some materials can carry electrical current without any energy losses. Measurements form the foundation of their search for answers. Once experiments and observations are made, a **hypothesis** (untested theory) is proposed and subsequently compared with additional data to test its validity. If a hypothesis successfully passes many tests, it takes on the status of a **theory**. This process is called the **scientific method** (**Figure 1.1**). Nevertheless, many advances in science came from simple trial and error experimentation or were simply a result of accidental discovery.

hypothesis Proposed untested explanation of an observation or experimental result.
theory A tested explanation of some natural phenomenon.

scientific method Process of asking and answering scientific hypotheses by making observations, doing experiments and evaluating them.

Scientific method Figure 1.1

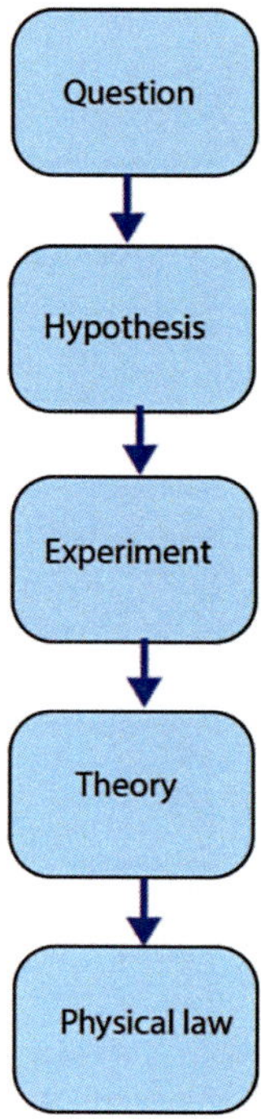

A theory leading to verifiable conclusions based on repeated scientific experiments and observations over many years becomes a **physical law**. Laws give science its predictive powers because they allow the knowledge of some input parameters values, such as the mass of an object and force on the object, to predict the value of an output parameter, such as how quickly the object changes speed (acceleration). We will study acceleration later in this chapter and examine this physical law, linking mass and force to acceleration, in Chapter 2.

physical law A theory leading to verifiable conclusions based on repeated scientific experiments and observations over many years.

However, theories and physical laws may not always be the last word on the behavior of nature because they are only our best and most recent descriptions. A next step that further improves upon and refines, or even refutes, current theory or law is always possible, if not always likely. A scientist needs to keep an open mind to be able to find and accept a revision to a long-established physical law that generations of scientists may have taken for granted. After all, laws and theories are only our best descriptions of our observations to date.

The pursuit of physical science follows along two tracks—experiment and theory. Experimentalists devise and perform experiments to explore new phenomena and also to test hypotheses. Theorists seek to develop new mathematical theories that agree with experiments and predict future results. They do this by plugging some input parameters into their equations and trusting the output of those same equations. Theory and experimentation depend upon each other. Progress in physics frequently comes about when experimentalists make a discovery that existing theories cannot explain or when new theories generate experimentally testable predictions, which inspire new experiments. If the theory's predictions cannot be measured, then the theory is not verifiable, and therefore, the theory is not useful. We say the theory's predictions are not testable.

Often, fundamental discoveries in physics that seem to have little immediate impact can lead to technological breakthroughs that improve our lives. For example, less than a year after Wilhelm Röntgen discovered x-rays in 1895, scientists found that x-rays could help to treat cancerous tumors. But the full potential of x-rays did not become apparent until the early 1970s, when scientists Sir Godfrey Hounsfield and Allan McLeod Cormack introduced an imaging technique known as x-ray computed tomography. This method—now known simply as CT scanning—made it possible to construct three-dimensional images of the human body for the first time.

Indeed, many of the technologies used in medical imaging and diagnosis depend on discoveries in physics that you will learn about in this book. Examples are magnetic resonance imaging (MRI), electrocardiography (ECG), electroencephalography (EEG), and sonography. Behind most devices that improve our lives lies a physics idea or concept of some kind.

Over the last 200 years, physics as a science has made enormous progress. Today the study of physics is divided into many different subfields, as shown in **Table 1.1**.

Table 1.1 Physics subfields

Mechanics The study of forces and the motion of objects.

Thermodynamics The study of temperature, heat, and energy.

Electricity and magnetism The study of electric and magnetic forces and currents.

Optics The study of light.

Atomic physics The study of the structure and behavior of atoms.

Nuclear physics The study of the nucleus of the atom.

Particle physics The study of the subatomic constituents of matter and energy, and the interactions between them. It is also called high-energy physics.

Condensed matter physics The study of the properties of matter in its solid and liquid states.

Astrophysics The physics of the universe, including the physical properties of celestial objects—such as galaxies, stars, planets, and the interstellar medium—as well as their interactions.

Standard Units of Measurement

Physics is founded on experimentation and observation. Objective observation is based upon the measurement of physical quantities such as length, time, and many others that you will learn throughout this book. The historical account of the evolution of different units of measurement provides many interesting stories and some amusing units. Many units of length measurement were based on body parts.

In ancient Egypt and Mesopotamia, people used their arms to define a cubit over 4,000 years ago. The cubit was the distance from the elbow to the tip of the longest finger. They further subdivided the cubit into smaller units called palms. Egyptian builders created the specifications for all their buildings, including the Great Pyramids, in cubits. An inch was originally defined as the distance between the knuckles of the thumb of King Edgar (A.D. 959-975), and the yard was proclaimed as the distance between the outstretched thumb and the nose of King Henry I (A.D. 1100-1135).

The Romans defined the first land mile as 1000 paces, with each pace equaling two marching steps. Marching their armies through uncharted territory, the Romans would often push a carved stick in the ground after each 1000 paces. In 1592, the English Parliament, during the reign of Elizabeth I, assigned the mile today's familiar 1,760 yards, or 5280 feet. We have not changed it since.

The existence of disparate measurement systems was one of the most frequent causes of disputes amongst merchants and between citizens and tax collectors. In 1670 a French vicar, Gabriel Mouton, proposed a *metric system* that used multiples of 10, 100, and 1000. His idea was slow to catch on; however, the system was later adopted in France in 1791, following the French Revolution. The French Academy defined the meter as one ten-millionth of a line running from the equator through Paris to the North Pole. The metric system was not an immediate success though. It was abolished by Napoleon in 1812 and reinstated in 1840. This system has been refined and gradually adapted by most of the world since. The precise measurement of time has been a challenge as well. Ancient civilizations used planetary movements through space to measure time and make a calendar. Even today we define the day, month, and year in terms of Earth's motion around the Sun and the Moon around Earth. See **Figure 1.3**.

Day and night were divided into 12 equal hours already by the Egyptians about 3000 years ago. No mechanical clock existed at that time. However, a sundial keeps time by tracking the motion of the Sun through the sky by the change of length and direction of an object's shadow. Sundials were used to measure time as early as 1500 B.C. Since Egyptians used arithmetic, which had 60 as its basis, they subdivided the hour into 60 minutes. However, the concept of a minute was impractical at that time because no precise clocks existed.

Rotation and revolution of the Earth around the Sun Figure 1.3

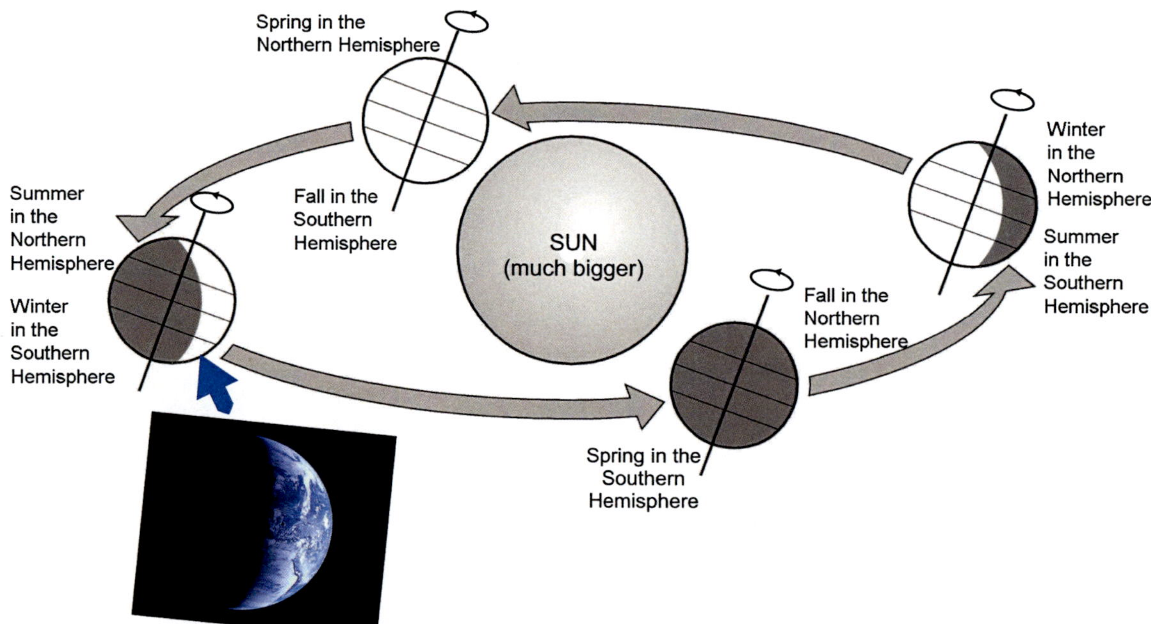

Early manual clocks used water, candles, or sand to measure the passage of time. However, none of these clocks could be left unattended. Not until the arrival of mechanical clocks in the 14th century did people begin using the minute as a unit of time, but clocks were rare and people usually guessed time as recently as the early 1800.

The metric system is the dominant measurement system around the world today. This system uses meters as units of length. Many metric conversions that rely on metric prefixes are shown in **Table 1.2**. For instance, the meter is equal to 100 of the smaller units called centimeters, and for larger distances we use the kilometer, which is 1000 meters.

In the United States and the United Kingdom, the British engineering system is primarily used. This system measures length in inches, feet, and miles. Both the British and the metric systems use the second as a unit of time. Some countries use both systems alongside each other to measure speed or length. We will use the metric units in this text. It is important to be able to convert from one system to the other if you plan to travel. **Appendix A** shows more conversion factors between metric and British units.

Table 1.2 Metric prefixes

Prefix	Meaning	Scientific Notation	In Words
Giga	1,000,000,000	10^9	billion
mega	1,000,000	10^6	million
kilo	1000	10^3	thousand
centi	0.01	10^{-2}	hundredth
milli	0.001	10^{-3}	thousandth
micro	0.000001	10^{-6}	millionth
nano	0.000000001	10^{-9}	billionth

Appendix A Units of measurement and their conversions

1 meter (m) = 100 centimeters = 1000 millimeters = 39.37 inches
1 yard = 0.914 meter
1 kilometer (km) = 1000 meters = 0.62 mile
1 foot = 12 inches = 0.305 meter
1 inch = 2.54 cm
1 mile = 5280 feet = 1609 meters
1 centimeter (cm) = 0.39 inch
1 gallon = 3.785 liters
1 pound (unit of weight) is equivalent to 0.453 kilogram (unit of mass)
1 kilogram = 1000 grams is equivalent to 2.2046 pounds
1 liter (l, or L) = 0.908 dry quart = 1.057 liquid quarts
1 (metric) ton = 1000 kilograms is equivalent to 2,204.62 pounds

1.2 Speed, Vectors, and Velocity

Many early philosophers thought about and described motion. At first they did not focus on measurement, but rather on logic. Later they progressed from general observations to making measurements of the observations. Recordings of the motion of planets and other astronomical objects were some of the earliest physics measurements. The key questions behind any motion are how fast the object in question is moving and how we should describe this motion.

To describe motion accurately, we first need to measure it relative to a **frame of reference.** For instance, you can judge the motion of a moving car relative to the road: The road is a frame of reference. On the other hand, if you are riding in an elevator in a building, then in the frame of reference of the elevator, you are at rest. However, in the frame of reference of the building, you are in motion. In any description of motion, we always specify the frame of reference. A mathematician could define a coordinate system, such as the mutually perpendicular *xyz* axes, as a frame of reference against which she could measure distance and speed.

frame of reference A system against which to measure the motion of any object.

Displacement, Velocity and Speed

It is interesting to see how the early thinkers viewed motion. For instance, in about 445 BCE, the Greek philosopher Zeno proposed the interesting idea illustrated in **Figure 1.5**—a paradox, because it suggested that no motion can be completed, which clearly is not the case! It took mathematicians over 2500 years to resolve this paradox by showing that the sum of an infinite number of steps can lead to a finite number.

Zeno's paradox Figure 1.5

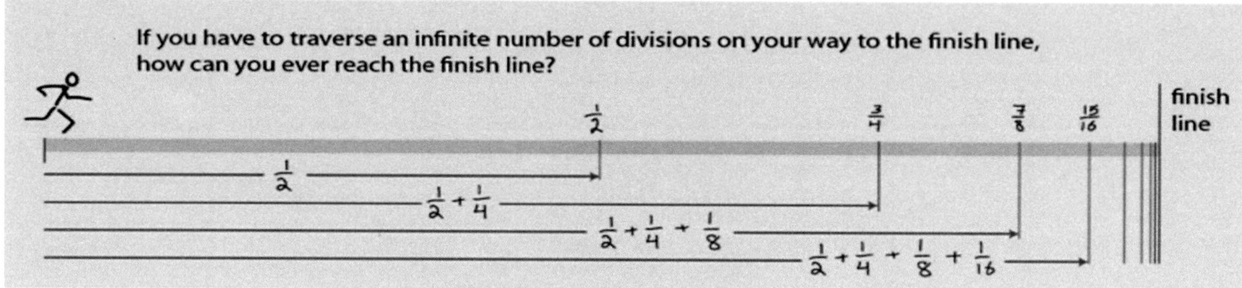

Any motion can be partly described by **speed.** We define average speed v_{av} as the **distance** traveled, *d,* divided by the time interval, Δt , taken traveling this distance.

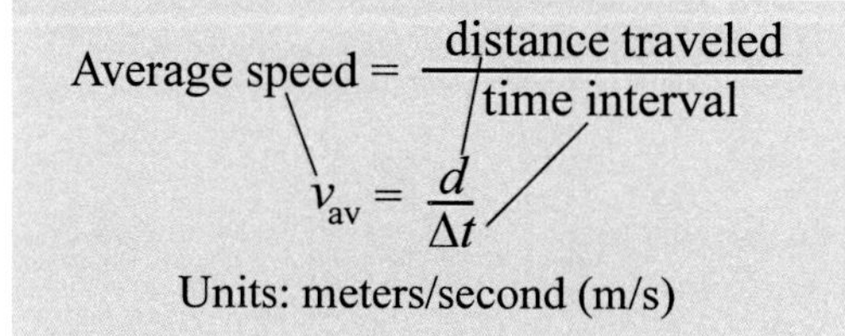

$$\text{Average speed} = \frac{\text{distance traveled}}{\text{time interval}}$$

$$v_{av} = \frac{d}{\Delta t}$$

Units: meters/second (m/s)

Here the Greek symbol Δ denotes "change in." An object's speed may vary during our measured time interval, so this relation gives us the average value only. If the object's speed varies, there will be time when the object travels faster than at the average speed and there will be time when it travels more slowly than at the average speed.

speed A measure of how fast an object is moving.

distance The length of the actual path traveled.

Example 1.2

Average speed of a car

A car travels for 2 hours at speed of 40 miles/hour, continues at speed of 50 miles/hour for 1 hour, and then at 70 miles/hour for 2 hours. What is the car's average speed? We assume here that the car changes the speed very quickly.

Solution:
To find the car's average speed, we determine the car's total distance traveled first. If the car travels at 40 miles/hour for 2 hours, we multiply the speed by the time traveled to determine the distance traveled or

$$\begin{aligned}\text{Distance traveled} &= \text{speed} \times \text{time interval}\\ &= \left(40\ \frac{\text{miles}}{\cancel{\text{hour}}}\right)\left(2\ \cancel{\text{hours}}\right)\\ &= 80\ \text{miles}\end{aligned}$$

Similarly, traveling at 50 miles/hour, the car travels 50 miles in 1 hour, and traveling at 70 miles/hour, the car travels 140 miles in 2 hours. The total distance traveled is 80 miles + 50 miles + 140 miles = 270 miles. The total time of travel is 2+1+2 hours, or 5 hours.

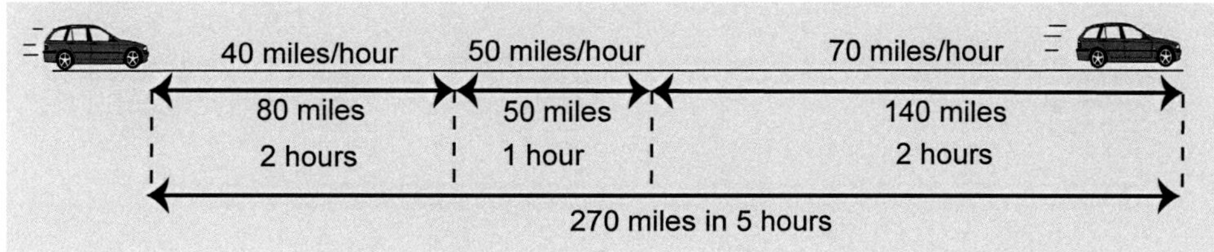

Now we are ready the find the average speed.

$$\begin{aligned}\text{Average speed} &= \frac{\text{distance traveled}}{\text{time interval}}\\ &= \frac{270\ \text{miles}}{5\ \text{hours}}\\ &= 54\ \text{miles/hour}\end{aligned}$$

Therefore, the car's average speed is 54 miles/hour. Note that the car travels at 54 miles/hour only at one instant as it increases its speed from 50 miles/hour to 70 miles/hour. The rest of the time the car's speed is either above or below its average speed.

Example 1.2 shows that if we know an object's average speed, we can determine the total distance it travels: It is the product of the average speed and the total time of travel. For instance, if you travel from St. Louis to Chicago in 5 hours at an average speed of 60 miles/hour, then you cover a total distance of 300 miles. We also learn that an object may travel at the average speed for only a brief time instant and the rest of the time it may travel at speeds either above or below its average speed. In that sense, average speed is just a calculated estimate of the object's speed over some time interval.

If we want the average speed to better reflect the actual speed, we can select shorter and shorter time intervals over which we compute this average speed. As we make the time interval arbitrarily small, we define an **instantaneous speed**, speed at one instant of time. If you are driving in a car and look at the car's speedometer to see how fast you are going, the number you see is your instantaneous speed.

instantaneous speed Speed at one instant of time.

Speed has many different units. We can measure speed in any combination of length units divided by time units, such as meters per second, kilometers per hour, feet per second, miles per hour, millimeters (milli = 10^{-3}) per second, nanometers (nano = 10^{-9}) per second and so on. **Application 1.1** discusses the speed of a glacier in units of kilometers per year, for instance. **Example 1.3** finds the average speed of a speedskier and shows how we can convert meter/second speed to an equivalent kilometer/hour speed.

Application 1.1 Speed of Glacier

A river of ice, located in the southwest of Greenland has become the fastest-moving glacier currently known in the world, a 2012 study suggests. The Jakobshavn Glacier - thought to have spawned the iceberg that sank the Titanic - is moving about four times faster than it was in the 1990s. It is the result of record ice melting in recent years. In the summer of 2012, the glacier reached a record speed of more than 17 km per year - more than 46 m per day. The increasing velocity of Jakobshavn means that the glacier is adding more and more ice to the ocean, contributing to sea-level rise.

Example 1.3
Measurement of a speedskier's speed

The author visited the 2003 Velocity Challenge in Sun Peaks, Canada to produce *Physics and Speedskiing* film for his physics class. Speedskier John "Mad Cow" Hembel from Aspen, Colorado, won this World Cup race. Officials measured his speed by placing timing photocells 50.00 meters apart at the steepest part of the bottom of Headwall Hill. The photocells recorded a time interval of 1.040 s between his timing zone entry and exit. What was Hembel's winning average speed in the timing zone?

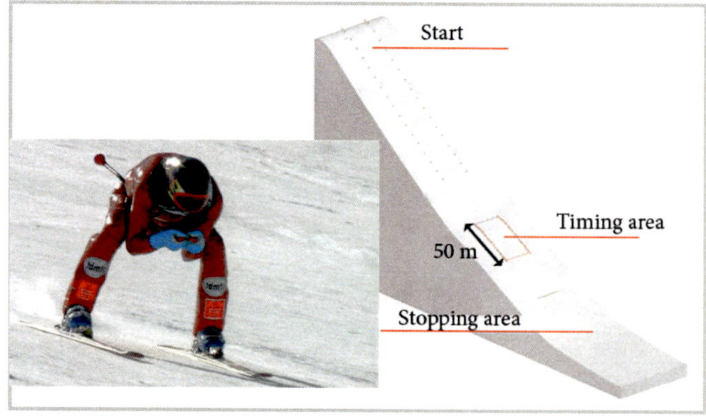

Solution:
We determine speed by dividing distance by time interval.

$$\text{Average speed} = \frac{\text{distance traveled}}{\text{time interval}} = \frac{50.00\text{ m}}{1.040\text{ s}} = 48.08\text{ m/s}$$

We can convert this speed to meters/hour by multiplying by 3600 seconds/hour.

$$\text{Average speed} = (48.08\frac{\text{m}}{\cancel{\text{s}}})(3600\frac{\cancel{\text{s}}}{\text{h}}) = 173{,}088\frac{\text{m}}{\text{h}} = 173.09\text{ km/h}$$

Hembel's winning speed was just over 173 km/h. While this is a significant speed, at a course in Les Arcs, France, speedskiers reach speeds of over 250 km/h. There are many YouTube videos showing skiers hurling themselves down glaciers at these incredible speeds. We will return to this example once we learn about displacement and acceleration in the next section.

Displacement and velocity as vectors

So far, we have not concerned ourselves with the direction of motion. The direction of travel is as important as how fast an object is moving. For instance, if you stand in the center of a frozen pond and start skating with a speed of 1 m/s, where will you be 10 seconds later? It is impossible to tell just from knowing your speed. You also need to specify a direction, such as north, for instance. We need a precise direction to define some physical quantities, such as velocity or force. We call these quantities **vectors**. We draw arrows to represent and visualize vectors. The length of the arrow represents the *magnitude* of the vector, and its orientation gives the direction of the vector (**Figure 1.6**). The starting point of a vector is known as the *tail* and the end point is known as the *head.*

Vector representation Figure 1.6

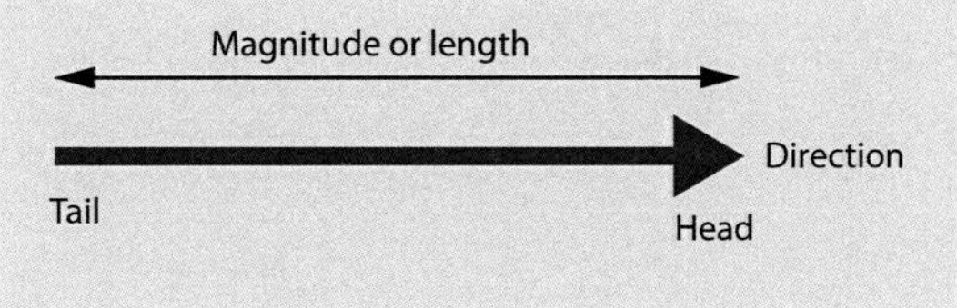

vector Physical quantity that has both a magnitude (or length) and direction.

On the other hand, a physical quantity that does not need direction to be uniquely determined is called a **scalar**. Speed is a scalar because no direction is associated with the concept of path length. Temperature is a scalar as well. We will study vectors here as this knowledge becomes useful later when we study physical quantities that depend on direction.

scalar Physical quantity that does not need a direction to be specified.

If an object can move only along a straight line, we say it's motion is one-dimensional. If we constrain this motion along a horizontal x-axis, a vector representing its motion can point only to the right or left and we can label the vectors as positive if they point to the right, or negative, if they point to the left. As you move from point A to point B, you cover some distance, but also you have **displacement** and those two may not be equal.

Displacement is a vector whose length is the shortest distance from the initial (point A) to the final (point B) position. On the other hand, distance is a scalar that refers to how much ground you covered during your motion. For instance, if you travel between two cities, displacement is the straight-line flight distance between the cities, while distance is the actual path your car took. Displacement can be positive or negative but the distance traveled is always positive. For an object that turns around, the distance traveled is greater than the distance between the end points.

displacement A vector pointing from the beginning point of motion to the ending point and whose magnitude represents the distance between these points.

Figure 1.7a and **Figure 1.7b** show the difference between displacement and distance for two different runs around your neighborhood. The total distance traveled is much larger than the magnitude of displacement in both cases. **Figure 1.7b** shows that for an 8 km run that starts and ends at your house, you cover a distance of 8 km, but your displacement from the start to the finish is zero.

Figure 1.7 Finding distance and displacement of a runner who finishes a) at different place than the starting point, and b) at the same starting point.

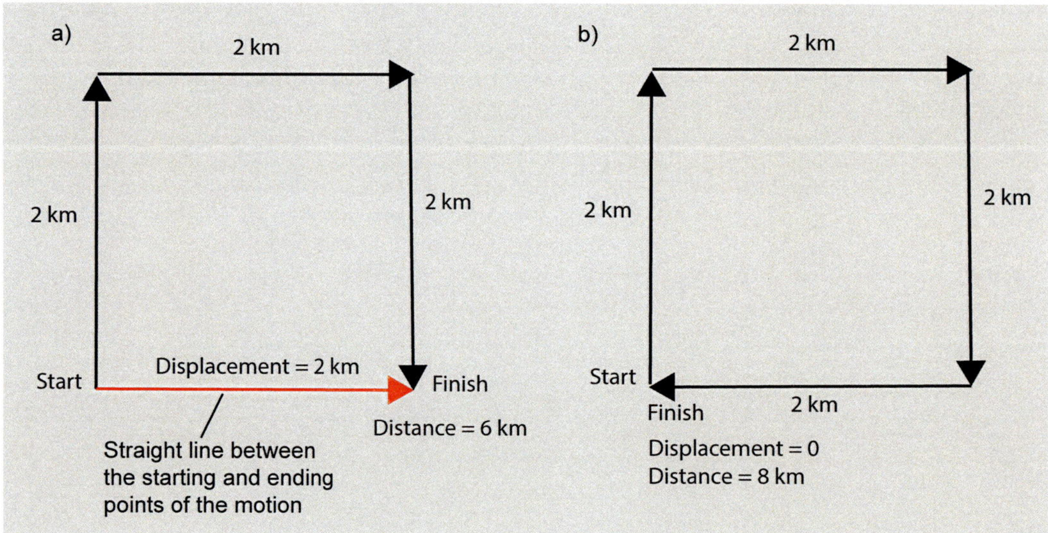

To fully characterize motion of an object we must also specify the direction in which it is moving. The "distance traveled" in the definition of average speed tells us nothing about the direction of travel. On the other hand, displacement gives us this information and therefore, we are ready to introduce a new vector quantity that tells us both how fast, *and* in which direction, an object is moving – **velocity**.

The definition of average velocity is similar to that of average speed, but here we replace distance with displacement. Average velocity, $\vec{v}_{av}$, is a vector defined as displacement divided by the time interval. To indicate that a symbol stands for a vector quantity, we put an arrow over it. As for speed, velocity measured over very small time interval is called *instantaneous velocity*. From here on, the word *velocity* without any adjective is meant to designate the instantaneous value of the velocity. Likewise, *speed* is meant to designate the instantaneous speed. When we are talking about the average speed or velocity, we always use the adjective *average*.

velocity Displacement divided by the time taken with the direction of motion defined.

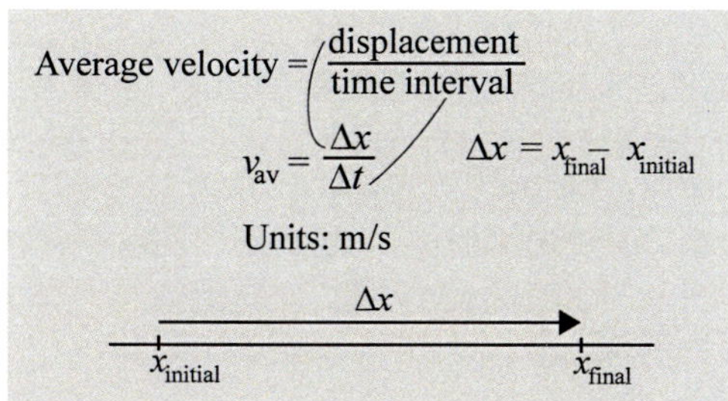

We chose to drop the vector notation here for 1-dimensional motion and use math signs to indicate direction. The average velocity takes its magnitude and units from the division of displacement and time interval quantities, but it takes its direction from the displacement alone. For motion in one dimension vectors are restricted to point only "right" or "left" for horizontal motion or "up" or "down" for vertical motion. For horizontal motion, the velocity is positive if the object moves to the right and it is negative if it moves to the left. For vertical motion, the velocity is positive if the object moves up and it is negative if the object moves down.

Speed is the magnitude of the velocity vector if the distance traveled matches the magnitude of displacement vector. This is always true if the time interval over which speed and velocities are measured is very small and we can say that *instantaneous speed is always the magnitude of instantaneous velocity*. However, the average speed may or may not be the magnitude of average velocity since the distance traveled may differ greatly from displacement.

Figure 1.8 gives a few examples of the vector nature of **velocity**. We can use standard compass directions (plus "up" and "down") to specify vector directions.

Drawing velocity vectors of cars traveling in different directions Figure 1.8

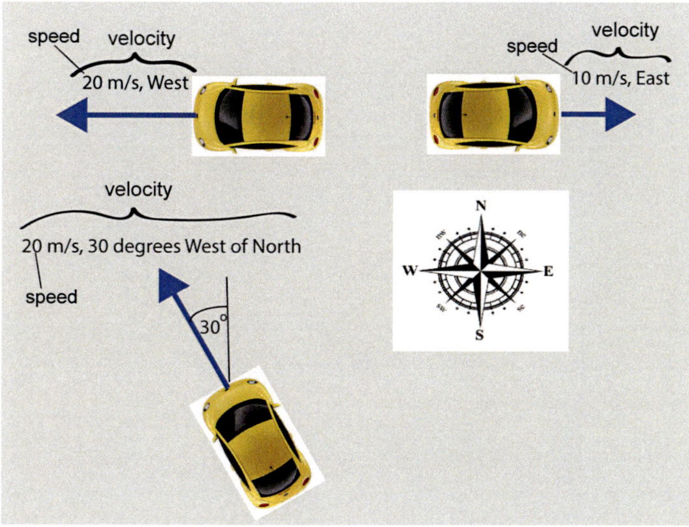

Velocity is a relative concept. For instance, when we say that a plane has a velocity of 600 km/h, heading west, we need to specify relative to what this velocity is measured. Is it measured relative to the ground or relative to the wind? If there is a strong tailwind or headwind, we may need to add or subtract two velocity vectors to find the net velocity relative to the ground. This procedure is called *vector addition* or *vector subtraction*.

We can add vectors graphically by drawing the first vector and then drawing the second vector with its tail at the head of the first one (**Figure 1.9**). Obtain the *resultant* vector by connecting the tail of the first vector with the head of the second vector. **Figure 1.9a** shows how to add vectors that are parallel to each other (1-dimensional vectors). In this figure, we also show what we mean by a negative vector and how we can treat vector subtraction as an addition of a negative vector. **Figure 1.9b** shows addition and subtraction of vectors in 2 dimensions.

Negative vector, and addition and subtraction of vectors that are parallel to each other Figure 1.9a

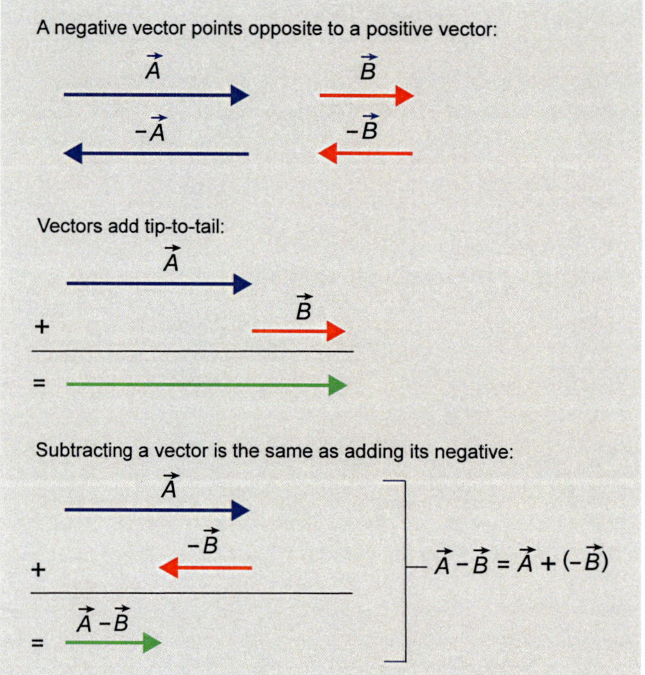

Addition and subtraction of vectors in 2 dimensions Figure 1.9b

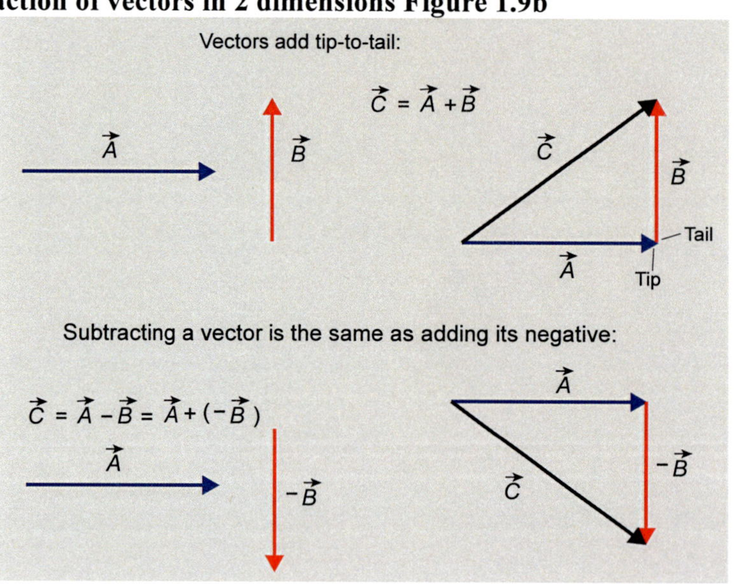

In one dimension, an object moving between positions x_1 and x_2 reached in corresponding times t_1 and t_2 has average velocity v_{ave}

$$v_{ave} = \frac{\Delta x}{\Delta t} = \frac{x_2 - x_1}{t_2 - t_1}$$

As the time interval approaches zero, we get instantaneous velocity $v(t)$

Graphing Motion

We can take a camera and capture the motion of an object such as a car at many different instants of time. Or we can sketch the position of the car at many different instants of time. Either way, this type of representation of the object's motion is called *motion diagram* (**Figure 1.10**). But drawing multiple images to represent motion is not very efficient and we learn to convert the motion diagram information, such as position and time, into a graph. See **Figure 1.10**. Position of a moving car is graphed on the vertical axis and time is graphed on the horizontal axis. The small red dots show positions of the car at different instants of time. The line connecting the dots helps us find the car's position at any other instant of time. The graph is an abstract representation of the motion.

How motion diagram is captured by a graph Figure 1.10

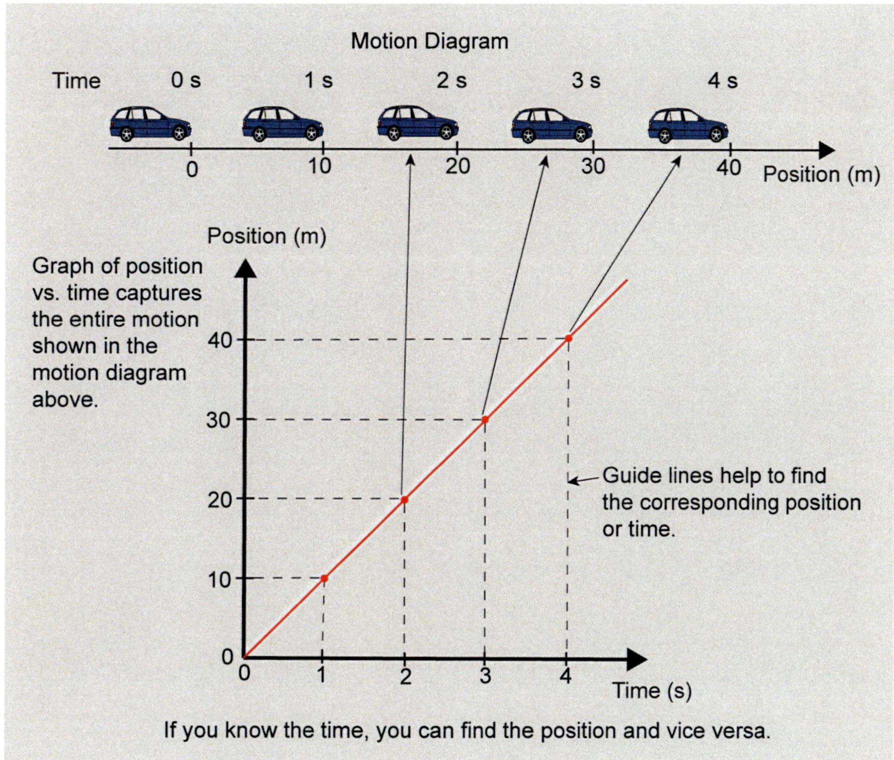

When an object moves with a constant speed, its graph on a position versus time graph is a straight line. Speed is the slope of this line. Remember that the slope of a straight line is defined as rise (change in position plotted on the vertical axis) over run (change in time plotted on the horizontal axis). See **Figure 1.11**.

Slope of the position vs. time graph is speed Figure 1.11

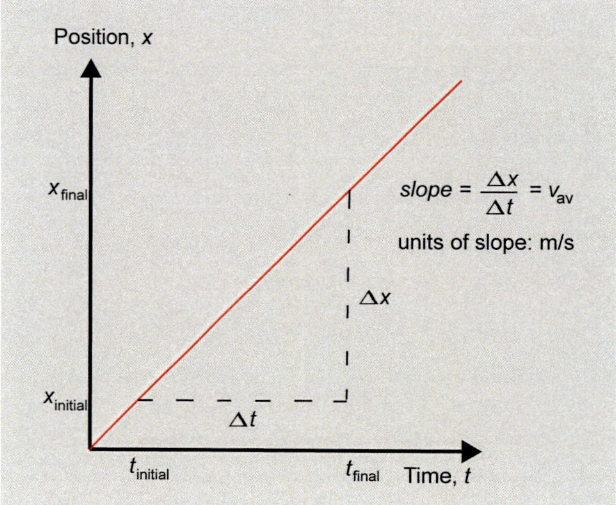

If the motion is along the x-axis only, then positive slope means the runner is moving forward (to the right) while negative slope means the runner moves backward (to the left). See **Figure 1.12**.

Graphing a round-trip run with different speeds Figure 1.12

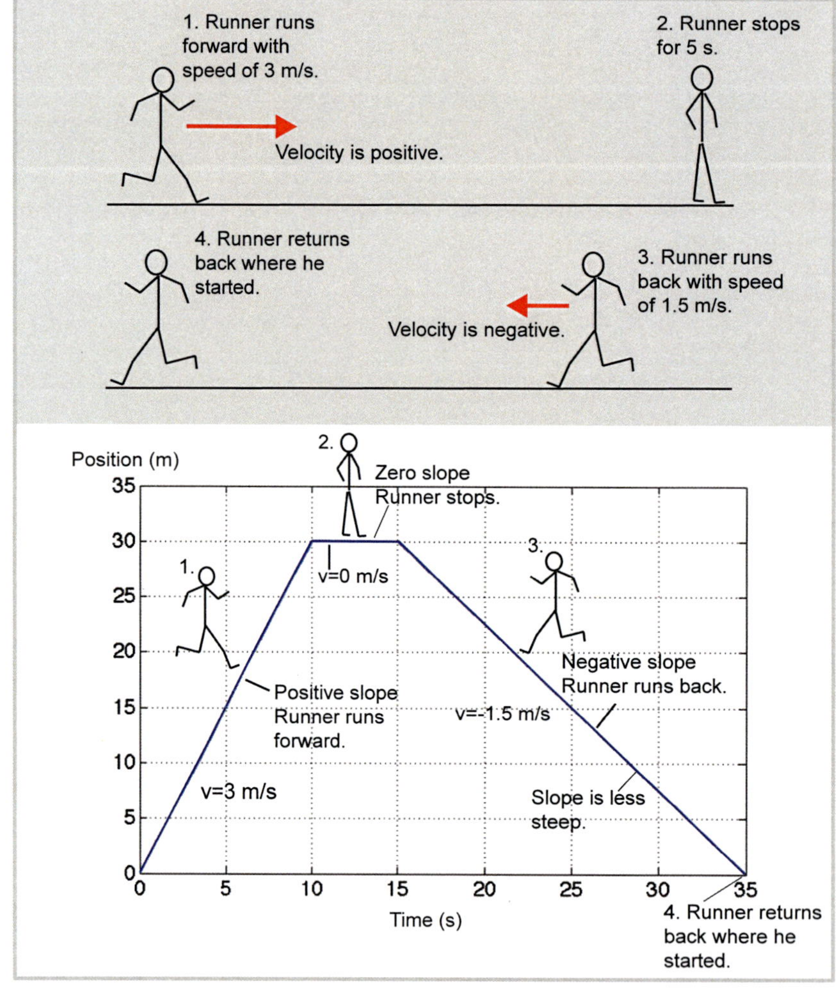

Typically, motion may not have constant speed. For instance, while driving in a car, we continuously speed up and slow down. A graph of position vs. time for such motion will be a more variable, or curvy line, rather than a straight line. Thus, the slope will be variable. You can determine the speed at some time instant from such a graph by measuring the slope over a tiny time interval. This slope will give you the instantaneous speed. The line on a position vs. time graph can slope upward or downward. This means the slope can have a positive sign or a negative sign.

We summarize the information that we can obtain from the position versus time graph:

1. We can find an object's position at any time by reading the graph at that instant of time.
2. We can find the object's average velocity over some time interval by finding the slope of the position vs. time line connecting this interval. Steeper slopes correspond to greater speeds.
3. We can find the direction of 1-dimensional motion by noting the sign of the slope. Positive slope means positive velocity and motion to the right (or up) and negative slope means negative velocity and motion to the left (or down).

Finally, straight-line motion in which equal displacements occur during any successive, equal time intervals is called *uniform motion*. This motion is also described as constant velocity motion.

1.3. Acceleration

We have seen how to describe velocity and focused on constant velocity motion so far. Any changes in velocity are caused by **acceleration**, our next topic of study. Acceleration describes change in velocity over some time interval; this change can be in magnitude or direction. For instance, speedskier John Hembel went from rest to about 48.1 m/s (173 km/h) in 5 seconds during his World Cup race (see **Example 1.3**). His increase in speed during those 5 seconds was dramatic. Here we will study how to define and measure acceleration such as his.

Acceleration as a Vector

If an object's velocity is changing, the object is accelerating. The determination of average acceleration $\vec{a}$ is similar to that of average velocity. The main difference is that now we measure change in velocity instead of change in position over some time interval.

$$\text{Average acceleration} = \frac{\text{change in velocity}}{\text{time interval}}$$

$$\vec{a} = \frac{\vec{v}_f - \vec{v}_i}{\Delta t}$$

final velocity ($\vec{v}_f$), initial velocity ($\vec{v}_i$), time interval (Δt)

Units: m/s^2

acceleration Change in velocity divided by the time it takes to make that change.

Acceleration is a vector. It has both magnitude and direction. If velocity is measured in meters per second, then the units of acceleration are meters per second per second, or m/s^2. In the above formula we can replace Δt by t if the initial time is defined as zero and the final time is t. In the speedskier example, John Hembel picked up speed and accelerated at a rate of (48.1 m/s)/(5.0 s) = 9.6 m/s each second, or 9.6 m/s^2.

For linear motion, a positive acceleration of 1 m/s^2 means that the magnitude of the velocity increases by 1 m/s each second. If the magnitude of the acceleration is constant, then we can also call it *uniform acceleration*. For a motion along a straight line, this means that the speed changes by the same amount each second.

For motion in 1-dimension, acceleration can be positive or negative, just like velocity. For instance, when you drive your car and step on the brakes, your car undergoes a negative acceleration, and the velocity and acceleration vectors point in opposite directions. See **Figure 1.13**. Here we use motion diagram by capturing each car image in equal time intervals. The car's displacement between each image is successively larger for positive acceleration and smaller for negative acceleration.

In general, for arbitrary velocity vector $\boldsymbol{v}$ we define average acceleration $\boldsymbol{a}_{\mathbf{ave}}$ as

$$\boldsymbol{a}_{ave} = \frac{\Delta \boldsymbol{v}}{\Delta t} = \frac{\boldsymbol{v}_2 - \boldsymbol{v}_1}{t_2 - t_1}$$

Acceleration and velocity vectors for positive and negative acceleration Figure 1.13

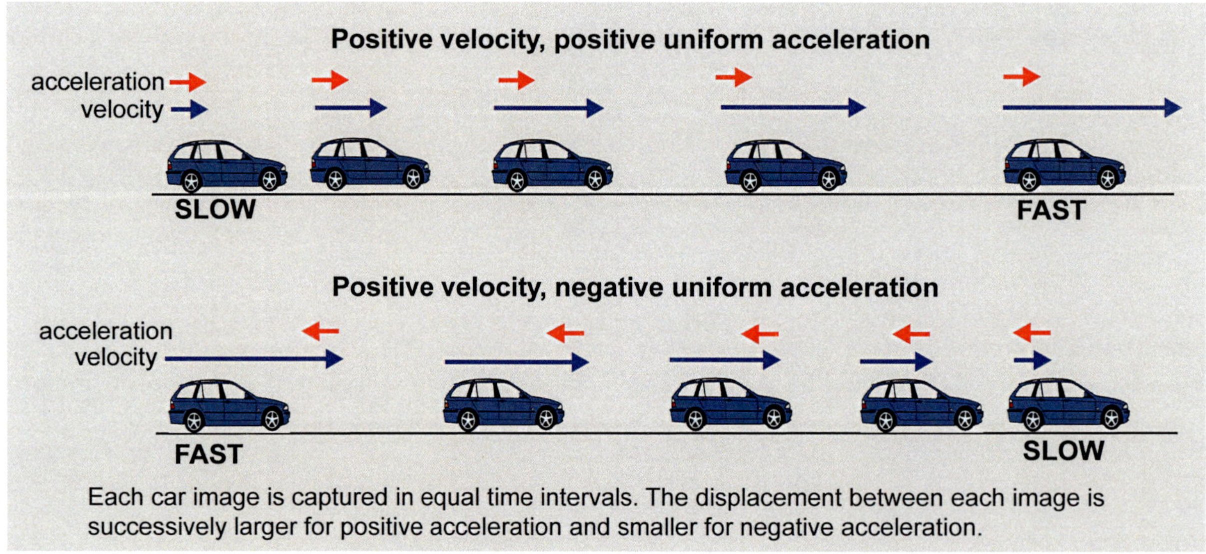

Each car image is captured in equal time intervals. The displacement between each image is successively larger for positive acceleration and smaller for negative acceleration.

Even if you move at constant speed, you can accelerate by changing your direction of motion. See **Figure 1.14**. Just like velocity, acceleration is a vector quantity. *Acceleration produces either a change in the speed or a change in the direction of your motion, or both.* For instance, as we will see in the next section, if you move around in a circle with a constant speed, you are still accelerating because your direction of travel is changing. Acceleration points in the direction of the change of velocity (**Figure 1.14**).

Object moving at constant speed but changing direction has acceleration Figure 1.14

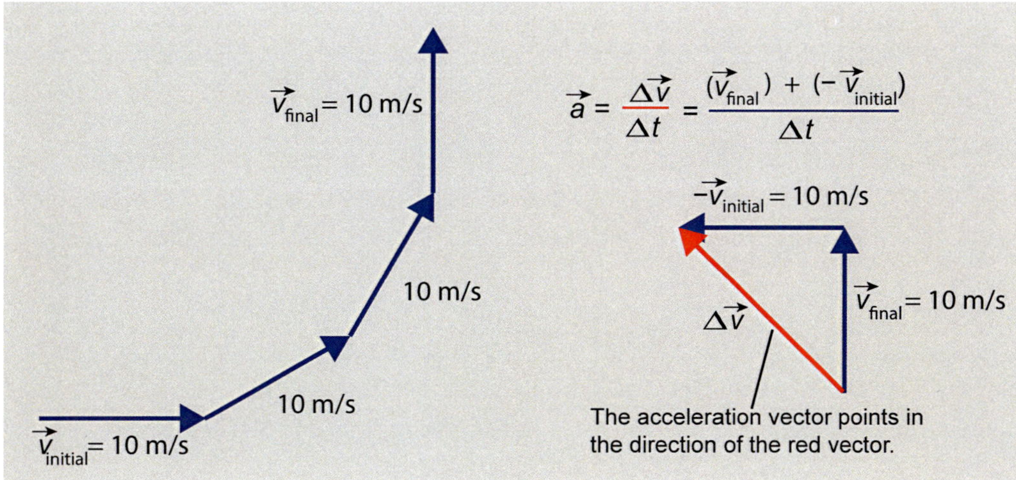

Knowing the acceleration of an object, we want to find the object's velocity at some instant of time later. We can rearrange the equation for average acceleration to find the final velocity $\vec{v}_f$ produced by the acceleration:

$$\text{Average acceleration} = \frac{\text{change in velocity}}{\text{time interval}}$$

$$\vec{a} = \frac{\Delta\vec{v}}{\Delta t} = \frac{\vec{v}_{final} - \vec{v}_{initial}}{t} \qquad t_{initial} = 0$$

$$\vec{a}t = \vec{v}_{final} - \vec{v}_{initial}$$

$$\vec{v}_{final} = \vec{v}_{initial} + \vec{a}t$$

Final velocity = initial velocity + acceleration × time

Here we are assuming that the initial time is zero and the final time is *t*.

Figure 1.15 shows the graph of this relationship for a car that has a constant (or uniform) acceleration––that is, a car whose speed increases at a constant rate. On a velocity-time graph, this motion is a straight line, indicating constant acceleration. The slope, or rise divided by run, of this line is the magnitude of the acceleration because the rise is the change in velocity and the run is the change in time.

Acceleration as the slope of a line on a velocity-time graph Figure 1.15

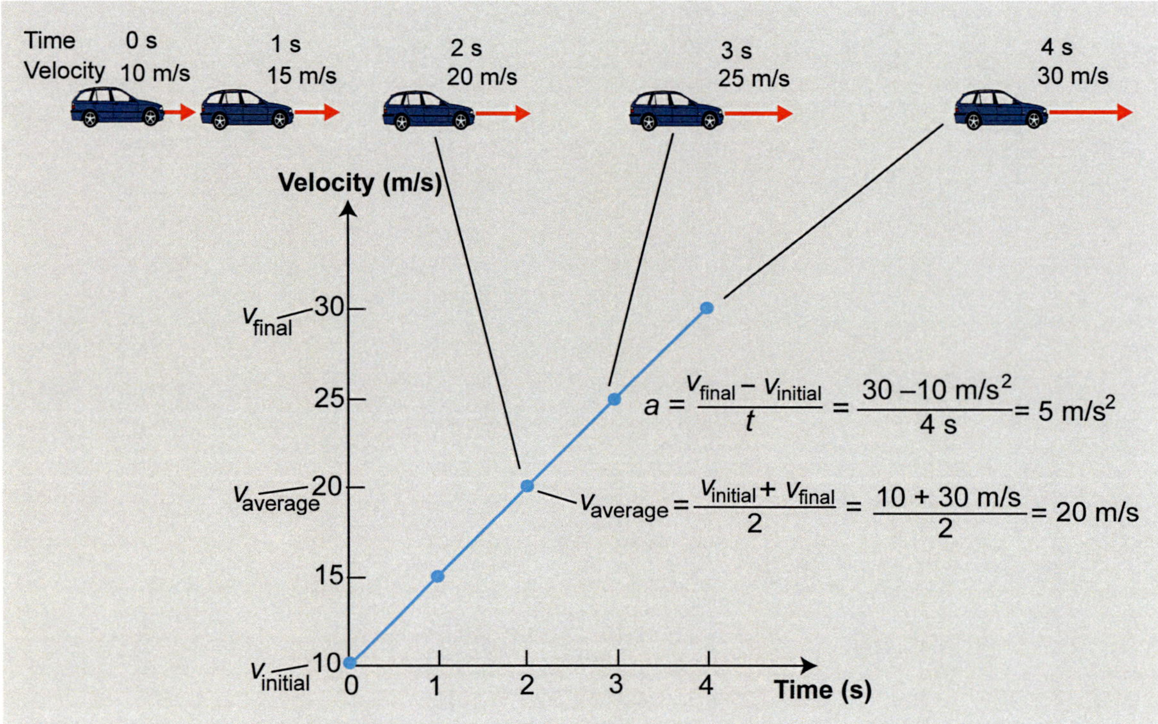

For a constant acceleration, where velocity is uniformly changing each second, the average velocity $\vec{v}_{average}$ is the midpoint of this velocity-time line. We determine the average velocity by adding the initial velocity $\vec{v}_{initial}$ to the final velocity $\vec{v}_{final}$ and dividing by two:

$$\text{Average velocity} = \frac{\text{initial velocity} + \text{final velocity}}{2}$$

$$\vec{v}_{average} = \frac{\vec{v}_{initial} + \vec{v}_{final}}{2}$$

We will drop the vector notation in our examples to simplify the appearance of the equations. This simplification is particularly convenient for 1-dimensional questions where object can move only along a straight line and the sign indicates direction.

So far we have studied how acceleration relates to initial and final velocities and the time of travel, but how about the distance covered during the acceleration? Note the separation between successive car images in the motion diagrams in **Figure 1.13** and **Figure 1.15**. The distance separation between any two successive images is larger than the distance between the previous images. During uniform acceleration, the velocity is continuously increasing and therefore the distance covered during any successive time intervals is increasing.

Recall, we found that skier John Hembel had an acceleration of 9.6 m/s^2 during the first 5 seconds of his run. We can find his displacement Δx by multiplying the average velocity by time, or $\Delta x = v_{av}t$. Since his initial velocity is zero, then the average velocity v_{av} is equal to half the final velocity v_f, or v_{av}=

$v_f/2$. Also, $v_f = at$, therefore $v_{av} = v_f/2 = at/2$. Multiplying this average velocity by time yields distance as a function of acceleration and time. See the full derivation below.

$$\Delta x = v_{ave}t = \left(\frac{v_0 + v}{2}\right)t$$

$$x - x_0 = \frac{1}{2}(v_0 + v)t = \frac{1}{2}(v_0 + v_0 + at)t$$

$$x - x_0 = v_0 t + \frac{1}{2}at^2$$

Thus, starting from rest, John Hembel traveled a distance of

$$\Delta x = v_0 t + \frac{1}{2}at^2 = 0 + 0.5\ (9.6\ \text{m/s}^2)(5.0\ \text{s})^2 = 120.0\ \text{meters}$$

during the first 5 seconds of his run. This relationship holds true for finding the distance traveled of any accelerating object. We see that for uniform acceleration distance traveled is proportional to time squared. Later in this chapter, we will show how Galileo first discovered this relationship experimentally while studying gravity. **Figure 1.16** visualizes how distance covered by uniformly accelerating puck increases with time. Notice that the change in distance for successive, equal time intervals increases as we expected. **Application 1.2** shows the dramatic change in speed during the landing of a space shuttle.

Change in distance covered by uniformly accelerating puck increases in equal successive time intervals Figure 1.16

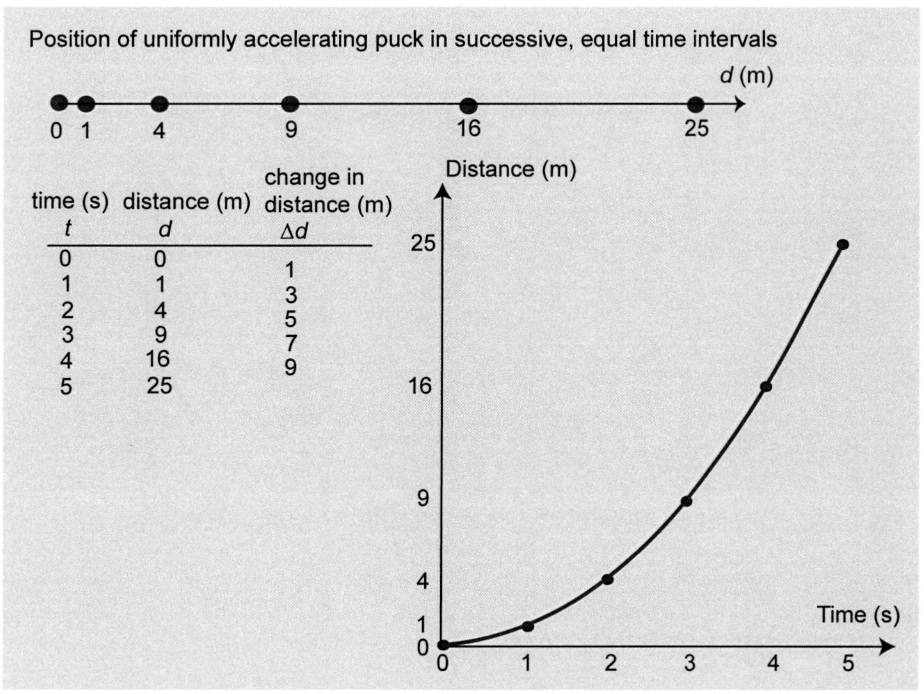

time (s) t	distance (m) d	change in distance (m) Δd
0	0	
		1
1	1	
		3
2	4	
		5
3	9	
		7
4	16	
		9
5	25	

Application 1.2 Negative acceleration during landing of space shuttle

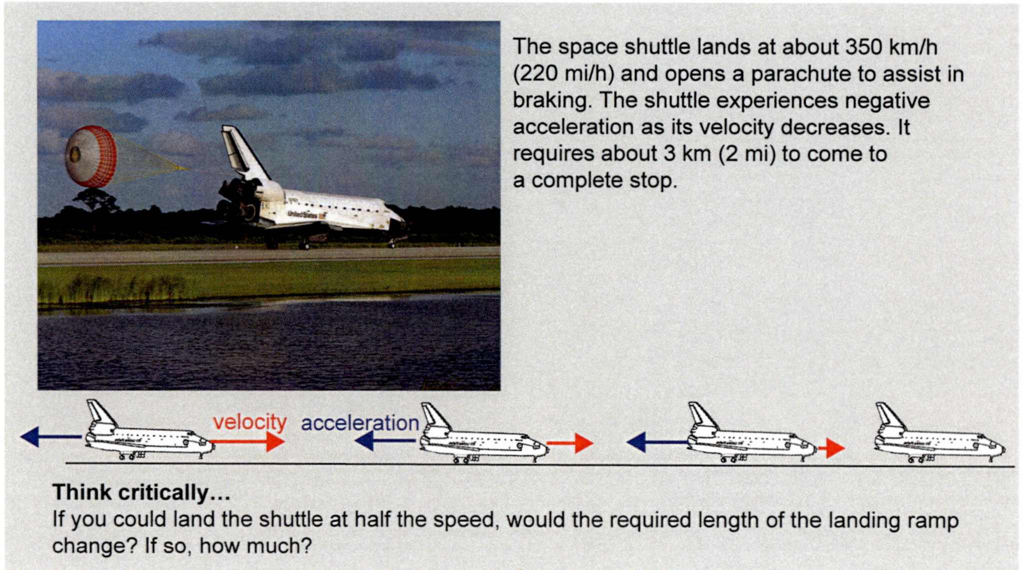

We are now ready to derive general quantitative relationships between displacement x, velocity v, acceleration a, and time t. Recall that average acceleration is a vector defined as the change in velocity over time taken to make that change or $a = \Delta v/\Delta t$. This leads to $v = v_0 + at$. Also, average velocity v_{ave} is given by

$$v_{ave} = \left(\frac{v_0 + v}{2}\right)$$

Then

$$x - x_0 = \Delta x = v_{ave}t = \frac{1}{2}(v_0 + v)t$$

But $v = v_0 + at$ and $t = \dfrac{v - v_0}{a}$

Therefore,

$$x - x_0 = \frac{1}{2}(v_0 + v)t = \frac{1}{2}(v_0 + v)\left(\frac{v - v_0}{a}\right) = \frac{1}{2a}\left(v^2 - v_0^2\right)$$

$$v^2 = v_0^2 + 2a(x - x_0)$$

This equation links initial and final velocities to displacement and acceleration.

Circular Motion

So far we have focused on acceleration as a change in the magnitude of velocity. Now let's study acceleration as a change in the direction of velocity. Any motion along a circle involves acceleration due to the changing direction of velocity, even if the magnitude of the velocity stays the same. This acceleration is known as **centripetal acceleration** (where "centripetal" is from the Latin for "center-seeking").

centripetal acceleration Acceleration directed toward the center of a circle for any constant speed circular motion.

If you whirl a ball on a string over your head, the ball undergoes centripetal acceleration. If you drive your car around in a circle, your car undergoes centripetal acceleration. A satellite orbiting the Earth also has a centripetal acceleration. For constant speed circular motion, the acceleration vector changes continuously and points towards the center of the circle. **Figure 1.17** demonstrates graphically why centripetal acceleration of a ball, moving in uniform circular motion, points toward the center of the circle.

The centripetal acceleration of an object moving along a circular path increases with velocity as velocity squared and increases inversely as the radius of the trajectory. The tighter the turn of an object, the higher is its centripetal acceleration. For an object moving with speed v along circular path of radius r, its centripetal acceleration a_c is given by

$$a_c = \frac{v^2}{r}$$

The centripetal acceleration always points toward the center of the circle and continuously changes direction as the object moves. **Figure 1.18** shows a proof.

Centripetal Acceleration Figure 1.17

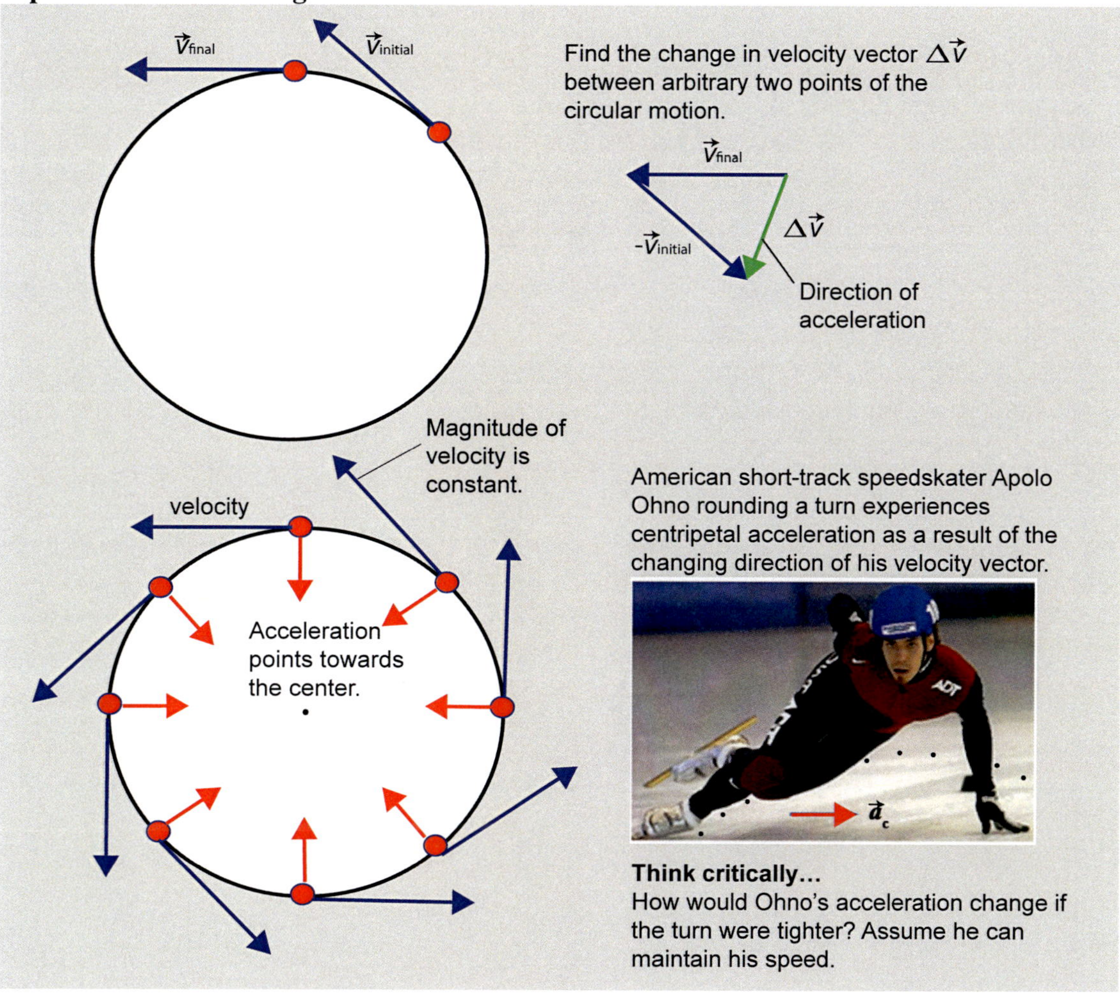

Derivation of centripetal acceleration Figure 1.18

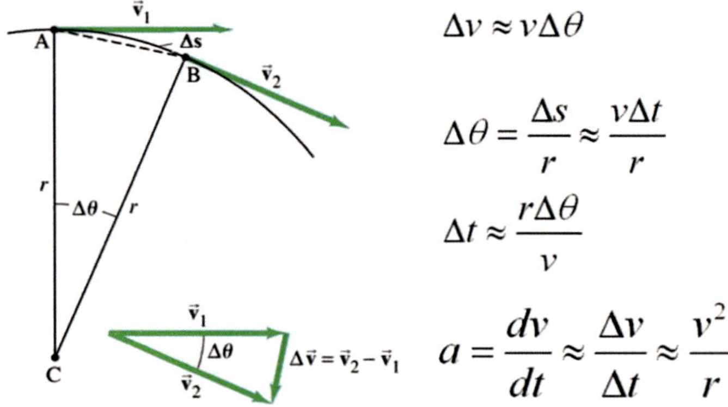

Acceleration Due to Gravity

If you hold an object in your hand and then drop it, it falls to Earth because of Earth's gravitational attraction. Any motion involving only the acceleration due to gravity is called *free fall*. The ancient Greek philosopher Aristotle (384 BCE – 322 BCE) was the first to try to understand free fall quantitatively. He made the assertion that heavier things fall faster. However, he did not check this assertion with an experiment because had he tested this hypothesis experimentally, he would have found that it is wrong.

Unfortunately, people accepted Aristotle's theory for centuries. Not until the early 17th century did Galileo Galilei disprove Aristotle's hypothesis. Aristotle was partially right, though. If you drop a feather, it will fall slower than, say, an apple or a bowling ball because of air resistance. Air resistance is negligible for a heavier object like a bowling ball, but it will slow down a feather considerably. However, if you were to place both the feather and the apple next to each other in a vacuum chamber, they would drop with the same acceleration and arrive at the bottom of the chamber at the same time (**Figure 1.19**).

The Moon does not have any atmosphere, and in 1969 Apollo 15 Commander Dave Scott stood on the Moon and confirmed Galileo's claim by dropping a hammer and a feather simultaneously. They landed at the same time, demonstrating that the weight of an object does not affect the time it takes to fall. The video of his experiment is available on the Internet. It is interesting to see how Galileo designed his experiment, finding a clever way to record time. He first dropped balls of different masses from great heights, but since they moved quickly and he did not have an accurate way of measuring brief spans of time, he could not make accurate observations. Instead, he "diluted" gravity by rolling balls down an inclined plane. He tried several ingenious methods to time the balls. One of them is shown in a replica of his inclined plane (**Figure 1.20**). He also timed the balls' descent with a water clock—a large vessel that emptied through a thin tube into a glass. After each run, Galileo weighed the water that flowed out—his measurement of elapsed time—and compared it with the distance the ball traveled.

Feather and apple drop with the same acceleration in vacuum chamber Figure 1.19

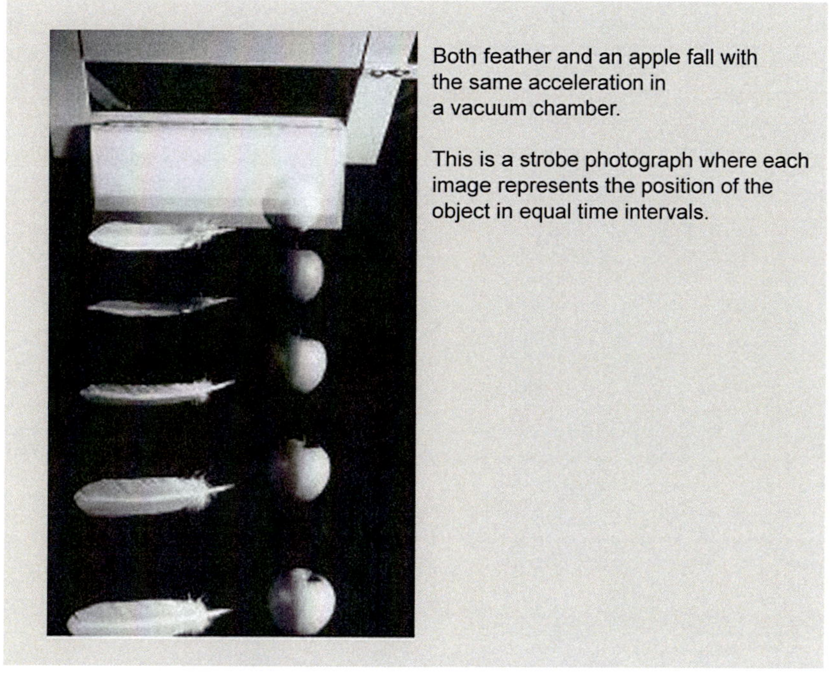
Both feather and an apple fall with the same acceleration in a vacuum chamber.

This is a strobe photograph where each image represents the position of the object in equal time intervals.

Galileo's measurement of acceleration due to gravity Figure 1.20

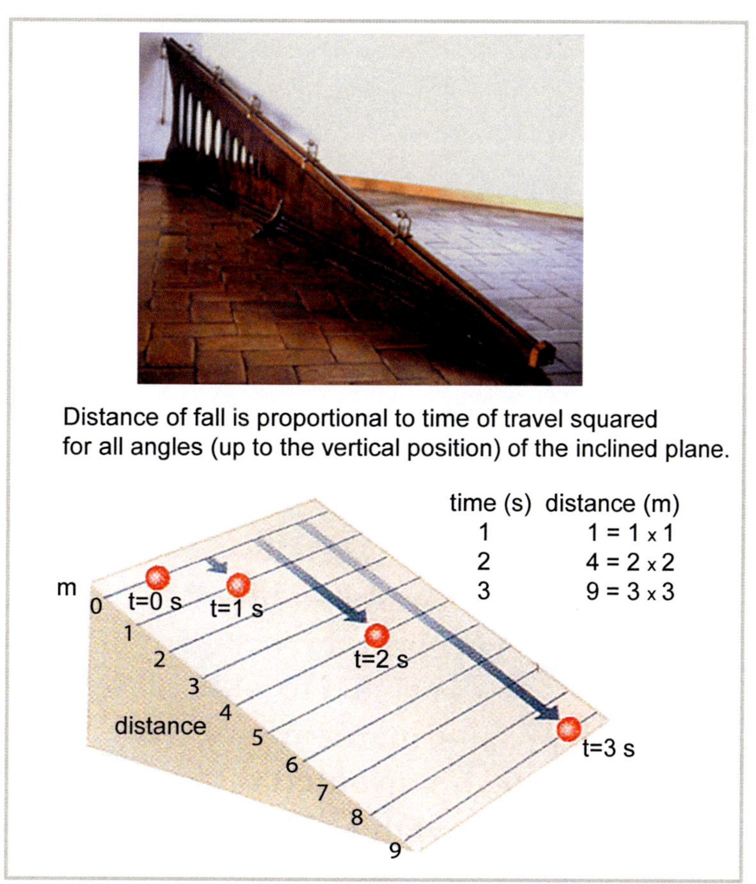

Galileo observed that in free fall, the *distance an object travels is proportional to the square of the elapsed time*. This is true for all angles of the inclined plane, all the way to a vertical fall (**Figure 1.20**). The numerical value of acceleration is higher for steeper inclines because the gravitational force (Chapters 2 and 3) along a steeper incline is larger. Galileo came to the following conclusion: *In the absence of air resistance, all objects on the surface of Earth fall with the same acceleration,* later precisely determined to be 9.8 m/s².

Figure 1.21a shows how velocity and distance of fall increase with time for a ball dropped from the top of a building. If the ball starts from rest and moves with constant acceleration a for time t, its final distance of fall is

$$d = \frac{1}{2}at^2 = \frac{1}{2}gt^2 = \frac{1}{2}(9.8\ \text{m/s}^2)t^2$$
$$d = 4.9\ t^2\ \text{m}$$

How velocity and distance of fall increase with time for a ball dropped from the top of a building
Figure 1.21a

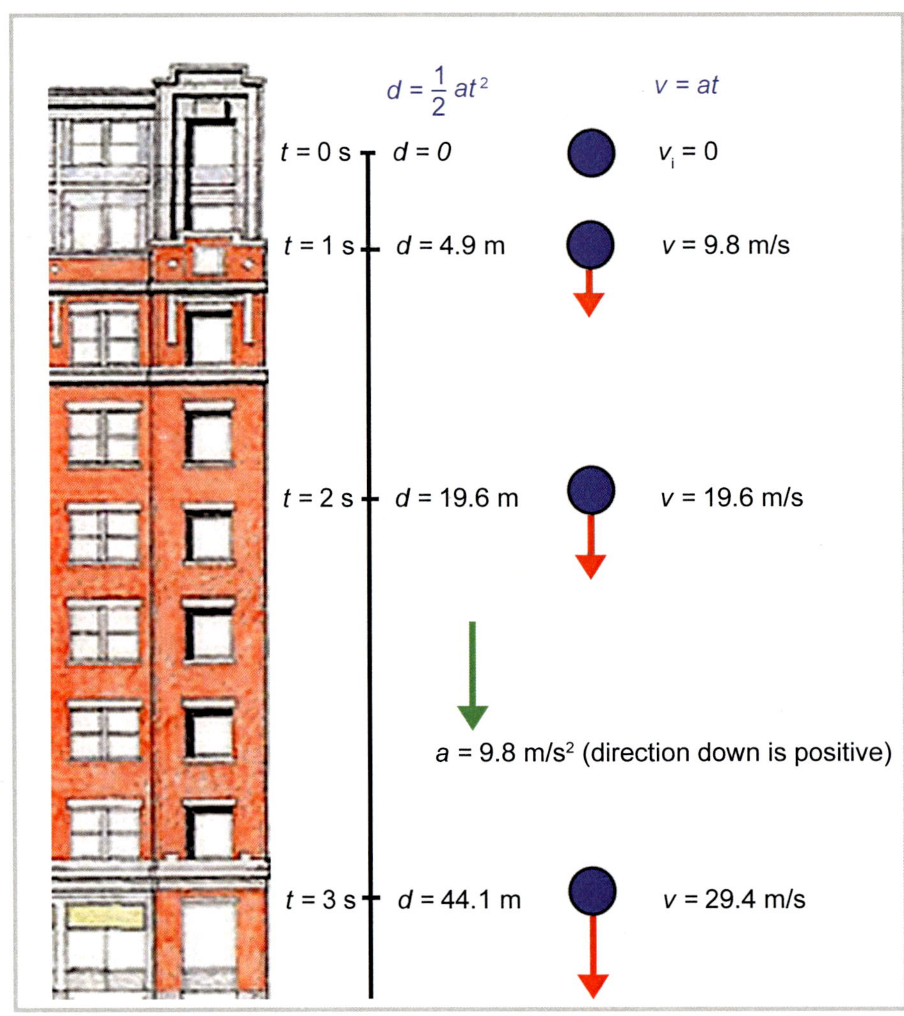

To find the ball's final velocity the general acceleration a is replaced by the acceleration due to gravity $g = 9.8$ m/s^2. The ball's final velocity magnitude, v_{final}, is

$$v_{final} = at = gt$$
$$v_{final} = 9.8\,t \text{ (m/s)}$$

Therefore, during free fall, distance of fall increases with the square of time, velocity increases directly with time, and acceleration remains constant.

This conclusion regarding the increase of the distance of travel with the square of time is true for any motion with constant acceleration. **Figure 1.21b** graphs the distance of fall of the ball in **Figure 1.21a** for different time intervals. This rapid increase of distance in time is also described as parabolic time dependence. We note that for one-dimensional travel where the object is constrained to move along a line and does not reverse the direction of travel, the displacement of the object and the distance of travel are identical.

Distance a ball falls after being released from the top of a building is proportional to time of fall squared Figure 1.21b

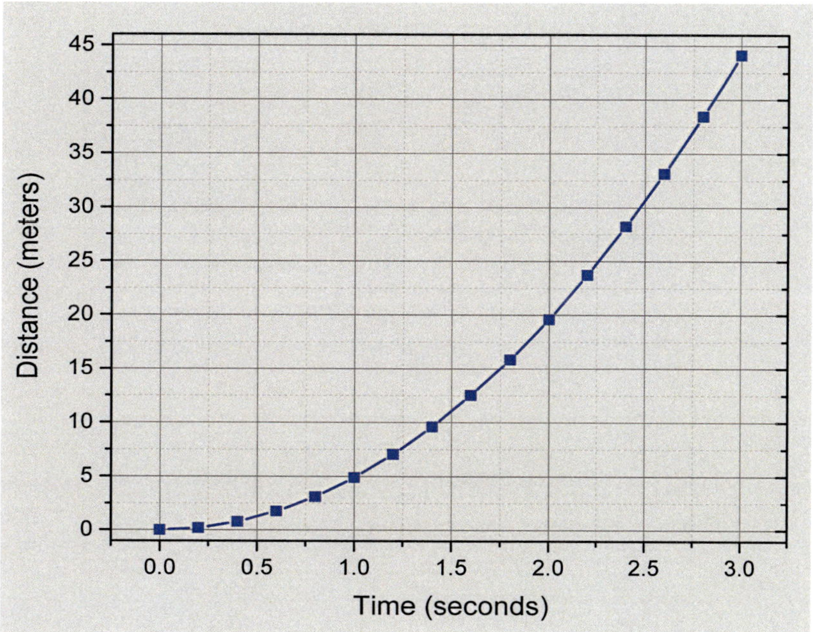

Example 1.7 describes the motion of a tossed ball.

Example 1.7
Motion of a tossed ball

You toss a ball straight up. Sketch a motion diagram so the images of the ball occur in equal and successive time intervals. Sketch velocity and acceleration vectors next to each image. Then sketch a velocity vs. time graph for the entire motion.

Solution: Once the ball is released its magnitude of velocity is continuously decreasing until it reaches the highest point of its motion or turning point. At that instant of time its velocity is zero. As the ball reverses its velocity, its magnitude of velocity starts increasing again. When the ball returns back to the release point, its magnitude of velocity is the same as when it was released. The motion is symmetric relative to the highest or turning point. On the way up, velocity is positive. On the way down, velocity is negative. At the highest point, velocity is zero.

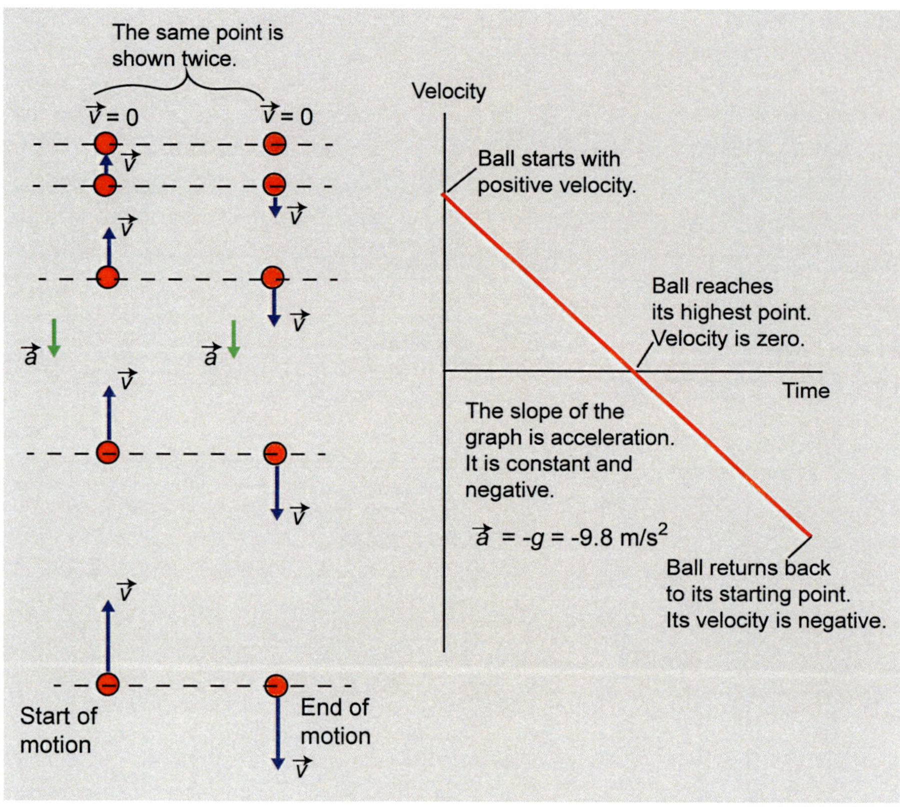

CONCEPT CHECK

1. **Why** can you accelerate even though your speed is constant?
2. **What** happens to the direction of centripetal acceleration during circular motion?
3. **How** can velocity and acceleration vectors point in opposite directions?

1.4 Projectile Motion

LEARNING OBJECTIVES

1. **Describe the** horizontal and vertical components of velocity and acceleration during projectile motion.
2. **Explain** any symmetry present in the projectile motion.
3. **Describe** the force needed to maintain circular motion and its dependence on speed and radius.

So far we studied the motion of an object such as a ball that is thrown straight up or straight down. We restricted our discussions to straight-line, one-dimensional motion. But what happens if you throw the ball out horizontally or at some angle? What is its trajectory like? Here we will look at two-dimensional motion confined to a horizontal or vertical plane. We will study **projectile motion** using examples like the path of the thrown ball, and we will also examine circular motion based on examples like the motion of a twirling ball or the path of the Earth on its annual journey around the Sun.

projectile motion Motion through space under the influence of gravity.

We have already learned that when we throw or launch something near the surface of the Earth, that object experiences a constant vertical acceleration due to gravity. We describe such motion as projectile motion. Already during late 16th century there was a strong interest in learning about projectile motion with the spread of canon warfare. Galileo performed a series of experiments to study projectile motion and his findings were in high demand throughout Europe. In one of his experiments he studied the motion of a ball moving down an inclined plane placed on a table and going off the table to the floor. He showed that the path of the ball, a projectile, follows a particular mathematical curve – a **parabola**.

parabola The curved path followed by a projectile under the influence of gravity only.

The term *parabola* describes the motion of a projectile when *only* the gravitational force is acting on the projectile. When other forces (such as air drag) are present, the path is no longer purely parabolic. For now, we assume that only gravitational force is present.

There are many examples of parabolic motion in the world around us. We can observe the fall of water droplets emitted by fountains or sparks from fireworks exploded in the sky, for instance. Sports give us a wealth of additional examples. Observe the paths followed by golf balls, tennis balls, or baseballs. They all follow parabolic trajectories. To better understand the two-dimensional nature of projectile motion, first let us examine vectors such as velocity that point at angles other than along the customary coordinate axes. We transform each vector into two parts, with each part being directed along one of the two coordinate axes.

For example, we can think of a vector directed northeast as having two parts: a northward part and an eastward part. Similarly, we can think of a vector directed southeast as having a southward part and an eastward part. These parts are called components of a vector. More specifically, we can call them the horizontal and vertical components of a vector (**Figure 1.22**). The combined influence of the two one-dimensional components is equivalent to the influence of the original two-dimensional vector.

Vector components Figure 1.22

We can describe a vector that does not follow along coordinate axes as being equivalent to two components that do.

a. Northeast vector has northward and eastward components.

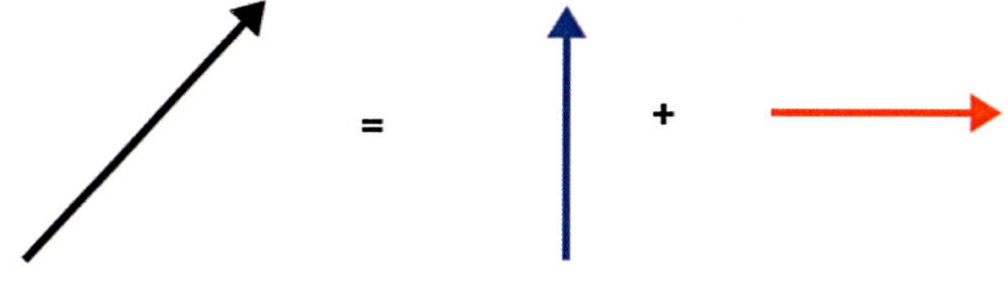

b. Southeast vector has southward and eastward components.

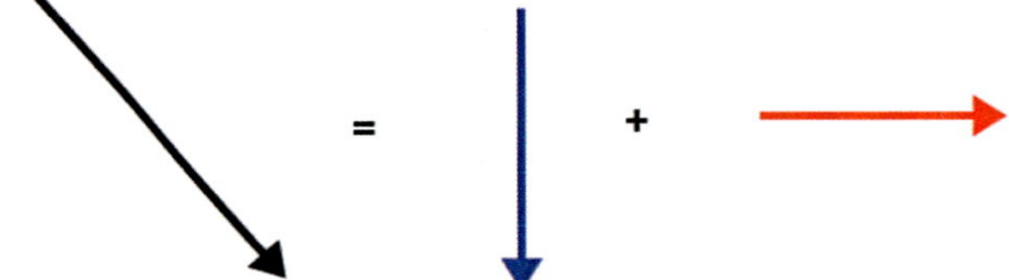

Projectile motion consists of two simultaneous motions: vertical and horizontal, described by horizontal and vertical velocity vectors. In the absence of air resistance, no horizontal force is present, so the horizontal motion has no acceleration. The vertical motion experiences a downward force of gravity. We can treat the horizontal and vertical motions independently of one another. The combination of horizontal and vertical motions produces a parabolic trajectory. See **Figure 1.23**.

Horizontal and vertical components of free fall Figure 1.23
One cannon ball is dropped and another is shot out horizontally at the same time.

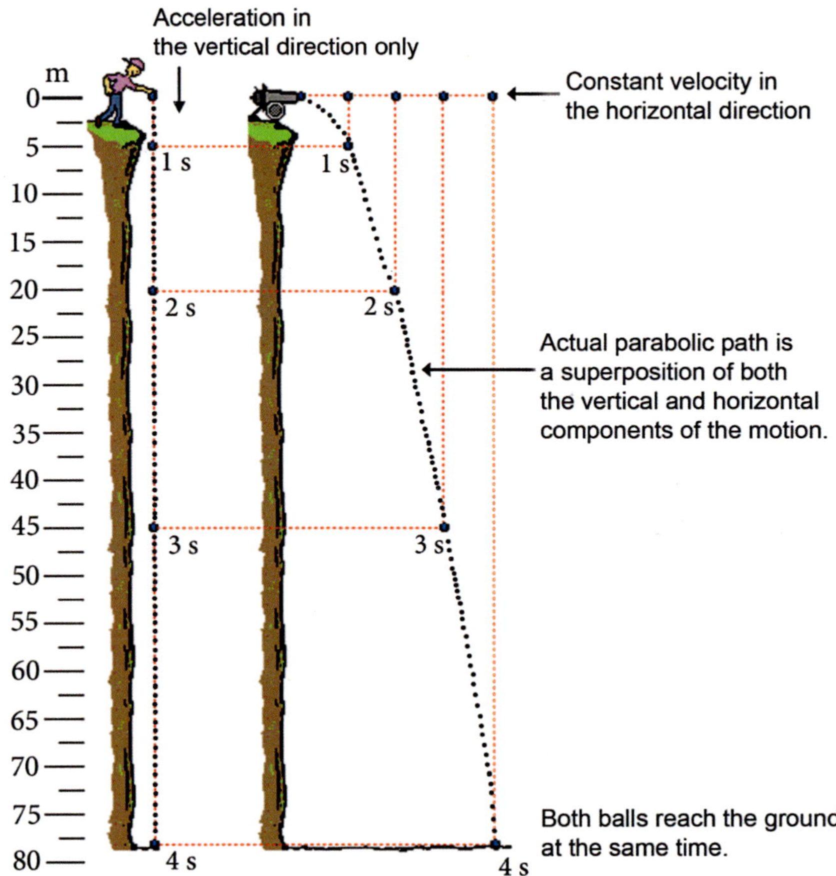

We can demonstrate the independence of the horizontal and vertical components of projectile motion by performing a monkey-dart experiment. Imagine chasing a runaway monkey with a tranquilizer dart gun. As the monkey is holding onto a tree limb, we take aim. At the instant we pull the trigger, the monkey lets go of the limb and starts falling. Will the dart hit the monkey? See **Figure 1.24** Since both the dart and the monkey fall the same distance in the amount of time it takes the dart to reach the monkey, the answer is yes. The dart falls at the same rate as the monkey. The fact that the dart is also moving horizontally does not change its rate of free fall: 9.8 m/s^2.

Dart aiming for the monkey experiment Figure 1.24
Both the dart and the monkey fall the same distance in the amount of time it takes the dart to reach the monkey.

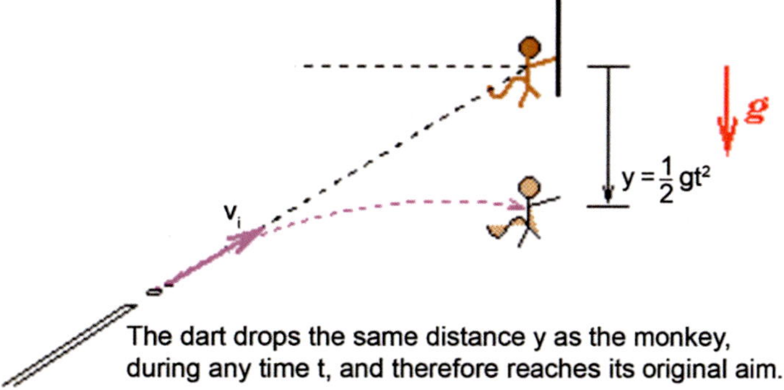

In the absence of air resistance, projectile motion is symmetrical (**Figure 1.25**).

Projectile motion and symmetry Figure 1.25
The horizontal component of velocity stays constant. The vertical component of velocity decreases as the ball approaches a peak (apex), where it is zero, and then increases again on its way down.

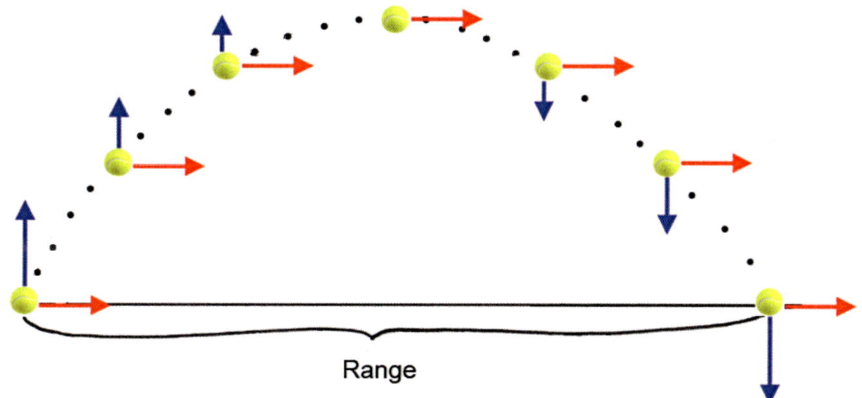

Projectile motion is symmetrical about a vertical line through the apex.

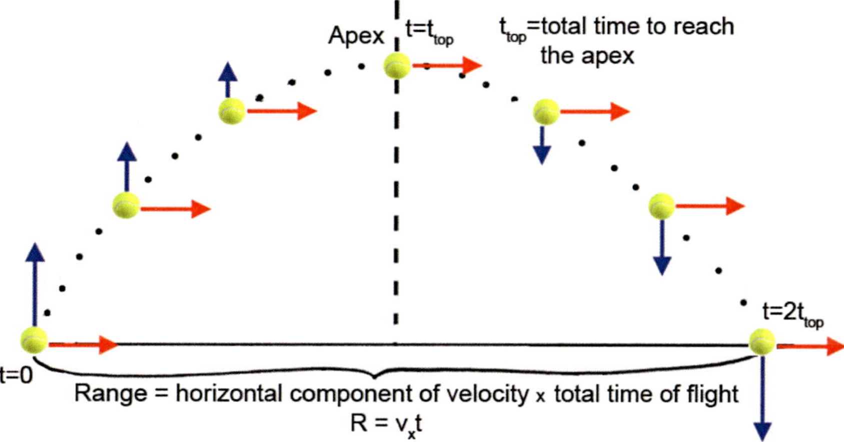

Time needed to reach the apex is the same as the time needed to return back to the ground.

If you drew a vertical line through the apex, you could fold the parabola along this central line and the half-parabolas would line up with each other. The vertical component of velocity flips direction after the projectile reaches the apex, while the horizontal component of velocity remains the same throughout its entire motion, since no gravitational force acts in the horizontal direction. As a result, time up equals time down.

You can determine the range, which is the horizontal distance of travel for projectile motion, by multiplying the time of travel by the horizontal component of velocity. Compare projectile motion both with and without the presence of gravity and the range of motion for different initial velocity angles. See **Figure 1.26**.

Projectile motion of a cannon ball with and without gravity Figure 1.26

a. Vertical drop of the cannon ball versus time of flight.

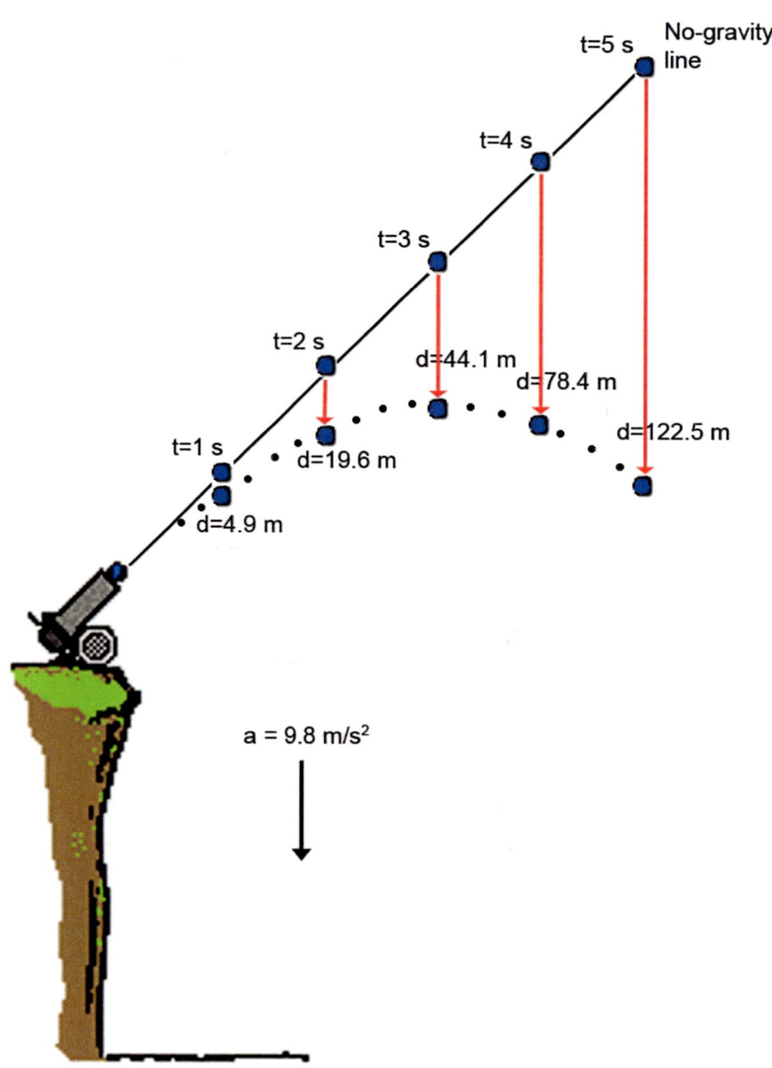

b. Note the gravity free straight line for both angled and horizontal initial velocity motion. The distance the ball falls during the same time interval is identical.

Time t	Vertical Drop $d = ½ a t^2$ where $a = 9.8 m/s^2$
1 s	$d = ½(9.8)(1)^2 = 4.9$ m
2 s	$d = ½(9.8)(2)^2 = 19.6$ m
3 s	$d = ½(9.8)(3)^2 = 44.1$ m
4 s	$d = ½(9.8)(4)^2 = 78.4$ m
5 s	$d = ½(9.8)(5)^2 = 122.5$ m

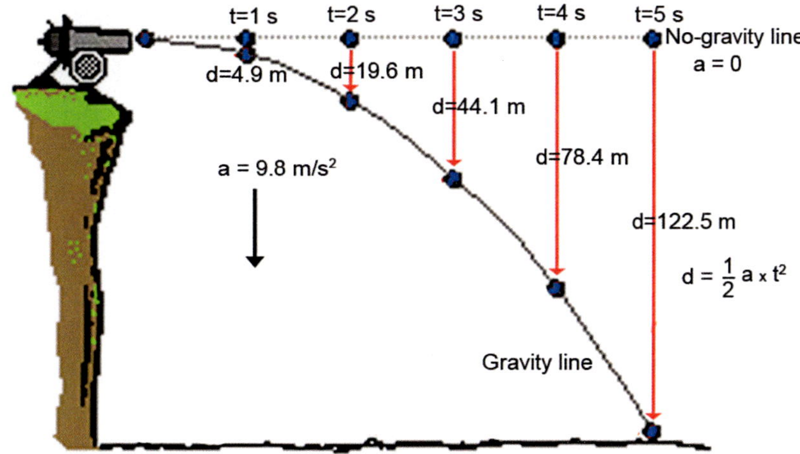

c. The range of the cannon ball shot at an angle of about 45° is longer than the range of the same cannon ball shot horizontally in with the same speed. In the absence of air resistance, shooting a cannon ball at an angle of 45° relative to the horizontal plane gives the maximum range

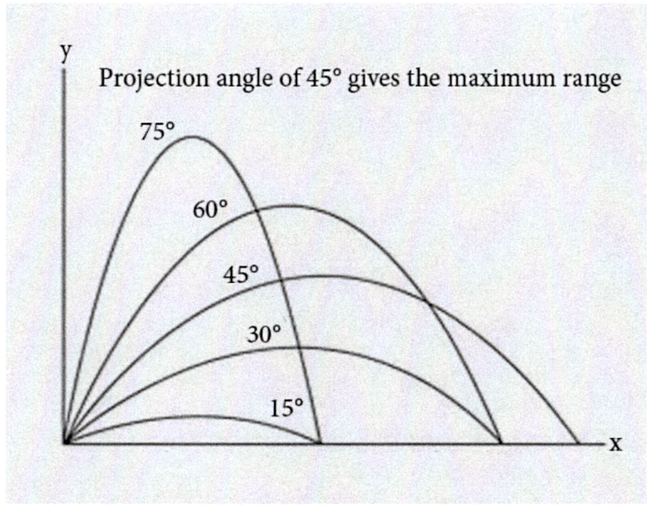

So far we have studied projectile motion in the absence of air drag. Air drag reduces the expected trajectory of a baseball or similar objects, such as a football, a tennis ball, or a golf ball. These changes are larger for faster moving objects. The trajectory is no longer parabolic or symmetrical. The moving object takes less time to rise and more time to fall back to the ground. The descent is steeper than the rise, and the launch angle for maximum horizontal range is usually several degrees less than 45° **(Figure 1.27)**.

Projectile motion with air drag Figure 1.27
The hitter's bat launches the baseball with a large initial velocity. Taking air drag into account, the ball travels a shorter distance, both vertically and horizontally.

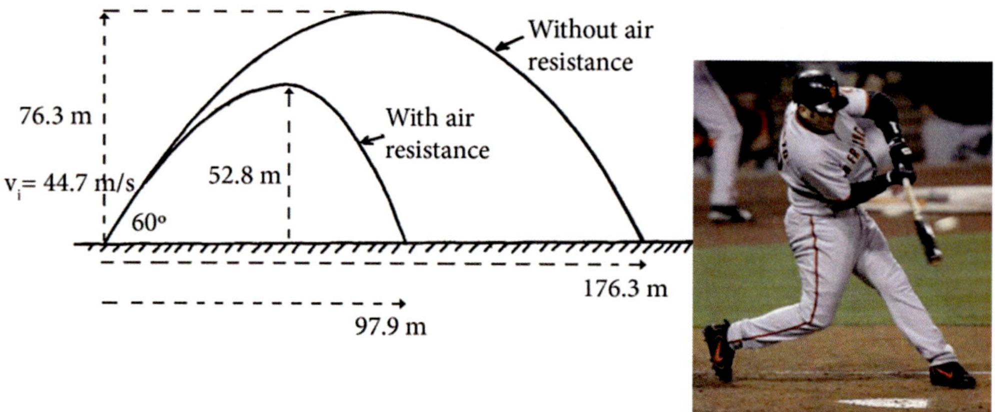

CONCEPT CHECK

1. **Which** ball, if either, would drop faster: one thrown horizontally, or one simply dropped down from the same height?
2. **What** is the difference between the time a thrown ball takes to reach the apex and the time it takes to come back down to the ground?

Summary

1.1 Physics and Measurement

- The **scientific method** is a process involving collection of data through observation and experimentation, and the formulation and testing of hypotheses.
- Most of the world uses one of two measurement systems: the Systeme International (SI), also known as the metric system, and the British engineering system, primarily used in the United States and the United Kingdom.

1.2 Speed, Vectors, and Velocity

- **Average speed** v_{av} is the distance traveled d divided by the time interval t spent traveling this distance. In symbols,

$$v_{av} = \frac{d}{t}$$

  The slope of a distance-versus-time graph gives us average speed.
- **Instantaneous speed** is speed at a given instant of time.
- A **scalar** is a physical quantity that is uniquely defined by its magnitude only. A **vector** is a physical quantity that has both magnitude and direction.
- **Displacement** $\boldsymbol{\Delta x}$ is a vector whose magnitude is the distance between the beginning and end points of the motion and whose direction is from the beginning point to the end point.
- **Average velocity** $\boldsymbol{v}_{av}$ is a vector defined as displacement $\boldsymbol{\Delta x}$ over time t, or

$$\boldsymbol{v}_{av} = \frac{\boldsymbol{\Delta x}}{t}$$

1.3 Acceleration

- Average **acceleration** is a vector defined as the change in velocity over the time interval to make that change, or
$$\boldsymbol{a} = \frac{\Delta \boldsymbol{v}}{t} = \frac{\boldsymbol{v}_{\mathrm{f}} - \boldsymbol{v}_{\mathrm{i}}}{t}$$
where $\Delta \boldsymbol{v}$ is the change in the velocity vector and t is the time interval.
- **Centripetal acceleration** is directed toward the center of a circle for any uniform circular motion.
- In the absence of air resistance, all objects on the surface of Earth fall with the same acceleration, equal to 9.8 m/s^2.
- During free fall, displacement d increases with the square of time t, velocity v_{f} increases directly with time, and acceleration g remains constant.
$$d = \frac{1}{2}at^2 = \frac{1}{2}gt^2 = \frac{1}{2}(9.8 \text{ m/s}^2)t^2$$
and
$$v_f = at = gt.$$

1.4 Projectile Motion

- **Projectile** motion is parabolic in the absence of air resistance. The motion can be treated as the superposition of separate horizontal and vertical components.

Key Terms

- **physics**
- **scientific method**
- **hypothesis**
- **theory**
- **physical law**
- **frame of reference**
- **speed**
- **instantaneous speed**
- **vector**
- **scalar**
- **distance**
- **displacement**
- **velocity**
- **acceleration**
- **centripetal acceleration**
- **projectile motion**
- **parabola**

Critical Thinking Questions

1. If the average velocity of a runner is zero, does that mean she is stationary? Explain.
2. Imagine that a book travels around the perimeter of a table of length 3 m and width 2 m and ends up in its original position. What is its distance of travel and its displacement?
3. You throw a ball straight up. Once it reaches its highest point, it briefly stops. What is its acceleration at that point?
4. If a car is moving eastward, can its acceleration be westward? Explain.

5. A motorbike racer in a turn maintains his speed but widens the radius of his turn. What happens to his acceleration?
6. You travel for two hours but your average velocity is zero. How can that happen?
7. When are you most aware of motion in a moving car? If the car moved with constant velocity (no bumps), would you be aware of the motion?
8. Can an object traveling with a constant speed be accelerating?
9. You stand in a moving train and drop a tennis ball. Describe two possible frames of reference the ball's speed will differ.
10. Does negative acceleration always mean that the object is slowing down? Can you think of a case where it is not true?
11. During uniform circular motion, the speed of an object is constant, yet the object has acceleration. Can you explain this apparent contradiction?
12. The acceleration due to gravity on the Moon is about one-sixth of that on the Earth. If you were standing on the Moon, and tossed a ball straight up with the same as on the Earth, describe the motion of the ball compared with that on the Earth. Will the ball rise lower or higher? Will the ball return sooner or later?
13. What would be the consequences if the gravitational acceleration doubled? How would it change any free fall motion?
14. If you were a scientist and suspected you might be on some new, undiscovered law of nature, how would you proceed to formulate it?
15. How would you measure "instantaneous" velocity? What is its relationship to average velocity?
16. When you subtract 2 equal scalars (for instance, mass), you get zero. But when you subtract 2 vectors, which have identical magnitudes, you get a range of possible answers, which depend on the orientation of the vectors. Identify the smallest and the largest possible answer and sketch it.
17. A student makes a statement: Two cars, A and B, are traveling with different speeds that they are not allowed to change. If car A overtakes and passes car B on the highway, while both cars are traveling in the same direction, then the cars have the same speed at the instant when B is alongside A. Do you agree or disagree with this statement and why?
18. A student makes a statement: Acceleration occurs only when an object is speeding up or slowing down. Do you agree or disagree with this statement and why?
19. In a uniform circular motion, the acceleration vector points towards the center of the circle. This is not very intuitive for the acceleration vector is perpendicular to the velocity vector. Can you explain, by making a small sketch, why the acceleration points towards the center?
20. If you had a strobe photograph, of an object moving on a long table in 1-dimension, how would you figure out if the object accelerating? If you were to compare several strobe photographs, how could you say which image shows larger acceleration?
21. Can you name some instances where the object's velocity is zero, yet has significant acceleration?
22. The graph of distance versus time below describes motion of an object. Determine the object's acceleration.

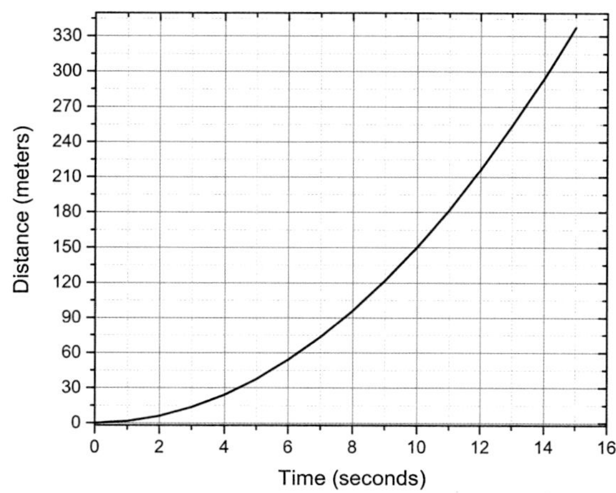

Exercises

1. At the 2009 World Track and Field Championships, Usain Bolt broke two world records by running the 100-meter dash in 9.58 seconds and the 200-meter dash in 19.19 seconds. What was his average speed in km/h and in mi/h?
2. A grizzly bear can run 500 meters in 30 seconds. What is its average speed? How would Usain Bolt (Exercise 1) do against him?
3. A hiker walks 5.0 km east and then 7.5 km west in 2.5 hours. What is her average speed and average velocity?
4. An ultra marathon runner's training route takes him 20 km north and then 32 km south. His run lasts 4.0 hours. What is his position at the end of the run? What is his total displacement? What is his average velocity? What is the total distance traveled? What is his average speed?
5. Imagine that an astronaut places a mirror on the Moon and you shoot a laser beam at it from Earth. The beam comes back 2.5 s later. How far is the Moon? Assume the speed of light is 300,000 km/s.
6. A shopper pushes a cart through a store. She moves forward 40 m, then she turns right and travels 20 m. What is the magnitude of her displacement?
7. A 1974 AMC Gremlin can go from 0 to 60 mph in 17.8 seconds, while the 1995 Ferrari 333 SP can do it in 3.6 seconds. What is their average acceleration in m/s^2?
8. If you accelerate from rest 1 m/s each second for an hour, how far will you be from your starting point?
9. You are standing on the edge of a cliff and drop a rock. How fast will the rock be moving 2 s later?
10. A helicopter hovers above a forest fire and dumps a large bucket of water. If you count off 3 seconds before the water lands, how high is the helicopter?
11. A car traveling westward at 25 m/s turns around and travels eastward at 15 m/s. If the turning around takes place in 5 s, what is the acceleration of the car?
12. A ferry is traveling straight across a river with a speed of 3 m/s north. The river's current has a speed of 3 m/s east. What is the ferry's resultant velocity as seen from the shore?
13. You throw a rock horizontally at a speed of 5 m/s from a bridge 50 m above a river. Ignoring air resistance, how long will it take for the rock to hit the water?
14. A cannon ball is shot horizontally from a cliff at a speed of 10 m/s. What is its speed one second later?
15. A snowboarder launches himself horizontally from the top edge of a cliff with an initial speed of 10 m/s. A stopwatch measures his drop time from top of the cliff to the bottom to be 4.4 s. How far out from the cliff's edge does he travel horizontally?

Chapter 2
Forces and Newton's Laws

What makes things move? This is an age-old question addressed by philosophers over many centuries, dating back to the early Greek thinker Aristotle and perhaps even earlier. Aristotle proclaimed that an object required a force in order to keep moving and that the natural state of an object was one of rest. But had Aristotle chosen to conduct some simple experiments to verify his hypotheses, he would had found that they were either incorrect or just partially correct.

Almost 2000 years later, Galileo Galilei (1564-1642) and especially Isaac Newton (1642-1727) helped explain motion and its causes by formulating theories that were verified by simple experiments and are accepted today. Newton laid the foundation of modern physics by introducing his three laws of motion together with the law of gravitation, which we will examine in the following chapter. These laws brought a sense of order and predictability to the universe and its interacting parts. In Chapter 1 we showed how to describe motion. In this chapter we will focus on Newton's laws as they explain the causes of motion and apply them to everyday examples, such as the process of pushing a car out of a rut in the sand.

2.1 Newton's First Law

LEARNING OBJECTIVES

1. **Describe** Newton's first law of motion.
2. **Explain** inertia of an object.

Galileo Galilei studied motion by doing a series of simple experiments. He observed motion by rolling a ball on a level surface and reasoned that if he could make a very long surface perfectly smooth, the ball would continue moving indefinitely. Therefore, contrary to Aristotle, Galileo concluded that objects could naturally remain in motion. However, he stopped there and did not address the causes of motion. It was Isaac Newton, born shortly before Galileo died in 1642, who took Galileo's ideas about motion a step further and used the concept of force to explain all motion.

Sir Isaac Newton (1642-1727) was born into a poor farming family. But farming did not interest him, and he went to Cambridge, England to study theology. Once at Cambridge, however, his focus turned to mathematics and physics, with great success. His studies led him to formulate the to become a preacher. However, once at Cambridge, he turned his focus to mathematics and physics with great success. Newton formulated laws of universal gravitation and motion—laws that explain how objects move on Earth and through the heavens. He also established the modern study of optics—the behavior of light—and built the first reflecting telescope. He helped develop a new area of mathematics called calculus.

Forces and Newton's First Law

Before we can examine what forces do, we need to understand what they are. Our commonsense idea of a **force** is that it is a *push* or *pull*. It helps to think of a force as a specific action: *forces act on objects*, just as pushes and pulls do. Every force has an agent, or an identifiable cause. For instance, when you throw a ball, your hand, while in contact with the ball, acts as the agent and cause of the force. A force can push or pull in any direction: it is a *vector*.

You cannot see forces themselves: you only see the changes they sometimes bring about. But note that not all forces cause changes in motion. For example, if you push on a large boulder, the force you apply will probably not be enough to move it.

When there are several forces acting on an object, we need to find the net or resultant force. We obtain the **net force** by adding all the force vectors acting on the object. It is possible to have two or more forces acting on an object but the net force is zero. Therefore, there is a difference between saying that no force is acting or no net force is acting on an object. Those two terms may not be the same.

force In the simplest sense, push or pull.

net force The vector sum of all forces acting on an object.

A common misconception dating back to the time of Aristotle holds that a moving body cannot maintain its motion without the application of force. Consider a hockey puck sliding across a table. Under normal conditions and with no additional applications of force, it will eventually come to a stop. But if we reduce the friction between the surfaces of contact (for example, by lubricating the puck), the puck will travel much further before stopping.

Now let us pretend that we could slide a hockey puck along an infinitely large frozen lake with a surface smooth enough to eliminate all friction. Under these conditions, the puck would slide along without any reduction in speed or change in direction. In other words, it would move forever in a straight line and at a constant speed. Galileo was the first to state that a moving object will continue moving in a straight line if no force acts upon it. Newton refined Galileo's idea and developed **Newton's first law of motion**. See **Figure 2.1**.

A puck moving at a constant velocity in the absence of a net force Figure 2.1
In the absence of a net force, the puck continues moving at a constant velocity indefinitely.

Newton's first law indicates that a state of rest (zero velocity) and a state of constant velocity are equivalent in that neither one requires a net force to sustain it. Only a net push or pull can change an object's speed or direction. It follows that *the application of a net force to an object does not sustain the object's velocity, but rather changes it,* as the example of riding in a turning car illustrates (**Figure 2.2**).

A net force changes your body's velocity in a turning car Figure 2.2
When you travel in a turning car, your body resists the turning action of the car because it has the tendency to continue its motion along a straight line.

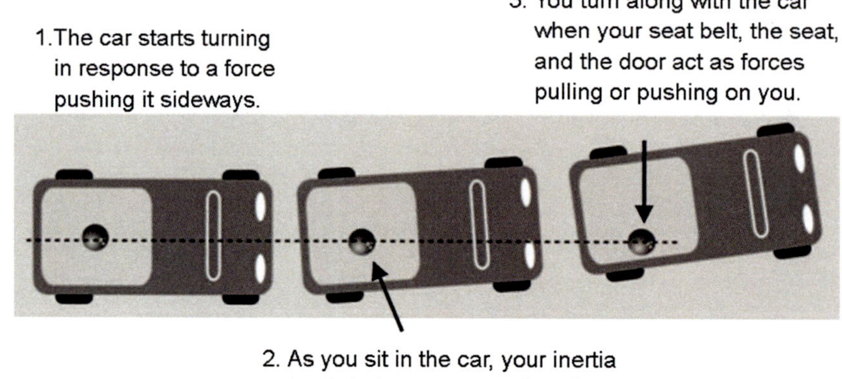

Newton's first law of motion Every object continues in a state of rest or in a state of motion at a constant velocity (constant speed in a constant direction), unless acted on by a net force.

Inertia and Mass

If you tried to move a car or other large object by pushing it, you would find that you needed a lot of force. The same would hold true of trying to change the direction of the car once it was moving. The property of objects that causes them to resist changes in motion is called **inertia.** It is the natural tendency of an object to remain at rest or in motion at a constant velocity. The concept of inertia relates directly to **mass.** Mass tells us the quantity of matter in an object. The larger an object's mass, the more difficult it is to move it: in other words, mass is a quantitative measure of an object's inertia. Its standard unit is the kilogram (kg). **Figure 2.3** shows how inertia helps a wet dog to dry off.

inertia The property of objects to resist changes in motion.

mass The measure of inertia.

Water's inertia allows a wet dog to dry off by shaking Figure 2.3

A wet dog shakes himself to dry off. His fur twists with the rest of his body, but the water droplets on his fur resist the change in motion because of inertia. The droplets tend to remain in place while the fur moves away from them, thus generating the spray that you see in the photo.

CONCEPT CHECK

1. **Why** would a puck sliding over a perfectly smooth frozen lake continue its motion indefinitely?
2. **What** happens to a necklace with a pendant hanging from the rearview mirror of a car that suddenly accelerates? Why?

2.2 Newton's Second Law

LEARNING OBJECTIVES

1. **Explain** Newton's second law.
2. **Describe** the difference between mass and weight.

In our initial study of motion, we defined acceleration as the time rate of the change of velocity. But what causes acceleration? Newton was the first to answer this question formally. His first law shows what happens when no net force acts on an object and no acceleration occurs. His second law shows what happens when a net force *does* act and changes the velocity as a result. The second law links net force to acceleration.

Force, Mass, and Acceleration

An important property of force is that it can change a body's state of motion. Hitting a tennis ball with a racquet changes its speed and direction. A racecar can accelerate to a high speed in a short time because its powerful engine can deliver a very large force. Whenever a net force changes the state of a body's motion, acceleration results.

Newton found and defined the relationship among force, mass, and acceleration. He concluded that...

1. The net force acting on an object produces acceleration in the same direction as the force. This acceleration is directly proportional to the magnitude of the force: the greater the net force, the greater the acceleration produced.

2. An object's acceleration resulting from a net force is inversely proportional to the mass of the object. The greater the mass of the object, the smaller the acceleration produced.

He summarized these observations in **Newton's second law of motion**:

$$\text{acceleration} = \frac{\text{net force}}{\text{mass}} \qquad \text{or} \qquad \text{net force} = \text{mass} \times \text{acceleration}$$

In symbols:

$$\boldsymbol{a} = \frac{\boldsymbol{F}_{\text{net}}}{m} \text{ or } \boldsymbol{F}_{\text{net}} = m\boldsymbol{a}$$

Newton's second law of motion The net force on an object is equal to its mass times its acceleration and points in the direction of the acceleration.

When the mass is expressed in kilograms and acceleration in m/s^2, the resulting unit of force is a newton, or N, in honor of Isaac Newton. The newton is a force that, when acting on a 1-kg mass, produces an acceleration of 1 m/s^2.
Example 2.1 applies the second law to the powerful motion of a baseball when hit by a baseball bat.

Example 2.1
Acceleration of a baseball
A major league baseball player strikes a baseball with an average force of 3,400 N. The mass of the baseball is 0.14 kg. What acceleration does he impart to the baseball?

Solution:
We apply Newton's second law of motion to determine the baseball's acceleration:

$$\text{acceleration} = \frac{\text{net force}}{\text{mass}} = \frac{3{,}400\,\text{N}}{0.14\,\text{kg}} = 24{,}286\frac{\text{m}}{\text{s}^2}$$

That is a huge acceleration!

Often several forces act on an object simultaneously. When that happens we need to find the vector sum of the forces to determine the net force. In the previous chapter, you learned to draw an arrow as a graphic representation of any vector, such as a force. The arrow's length should be proportional to the magnitude of the force, and the arrow should point in the direction of the force. Sum up the forces by following the rules of vector addition: add the tail of the second force vector to the tip of the first force vector. **Figure 2.4a** shows two tugboats pulling a ship in different directions. We can use the tip-to-tail method to find the net force acting on the ship (**Figure 2.4b**).

The second law tells us that a change in speed or direction will occur only if a net force acts on the object. When the net force is zero, the second law leads us to the first law. Zero net force gives us zero acceleration, which leads to constant velocity or a state of rest.

Net force acting on a ship Figure 2.4

a. Two tugboats pull a ship in two different directions, exerting forces T_1 and T_2. The combined net force is F.

b. Tip-to-tail vector addition of two forces, T_1 and T_2

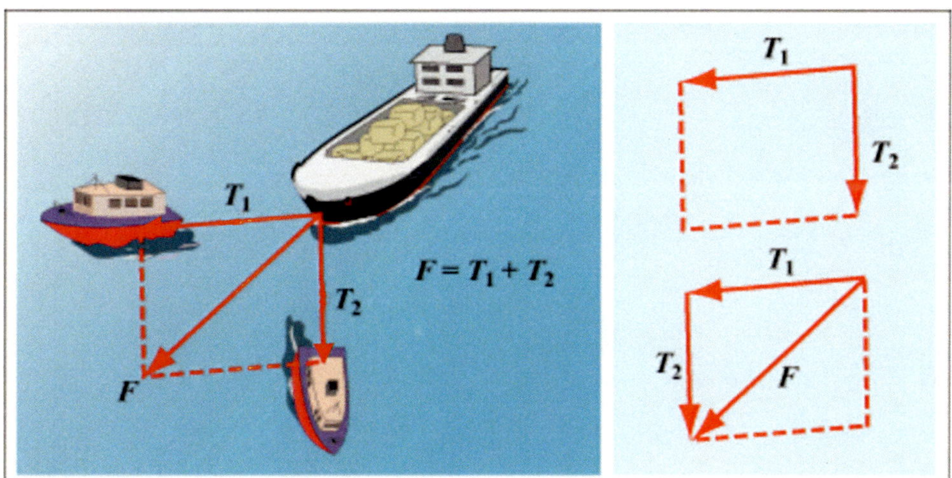

Weight

While there are many forces all around us, perhaps nothing is so ingrained in our senses as the Earth's pull on all of our surroundings. We know that if we release an object it will fall to the ground. This is because an attractive gravitational force exists between the object and the Earth. This gravitational pull on a body is its **weight**, which we can calculate using Newton's second law, $F = ma$. The acceleration due to gravity at the Earth's surface is $g = 9.8\ \text{m/s}^2$. Substituting g for acceleration, we obtain:

$$\text{Weight} = \text{mass} \times \text{acceleration due to gravity}$$

In symbols,

$$w = mg$$

weight Gravitational force equal to mass times acceleration due to gravity.

Since the acceleration due to gravity is approximately constant on the surface of the Earth, people often use the terms mass and weight interchangeably. But they are different. *While mass will always stay the same, weight will change with gravitational acceleration*. For instance, the acceleration due to gravity on the Moon is approximately six times smaller than that on the Earth. Thus the weight of an object on the Moon will be about one-sixth of what it is here on the Earth, and in deep space, away from any massive objects, it could be zero.

Example 2.2 shows how we apply Newton's second law to a question that involves weight of an object.

Example 2.2
Sliding block on a pulley
A block of mass M sits on a table and is attached to a rope that passes over a pulley near the edge of the table and another block of mass m, suspended by the rope, hangs in air. If the table's surface is frictionless, mass m will move downward. Find its acceleration.

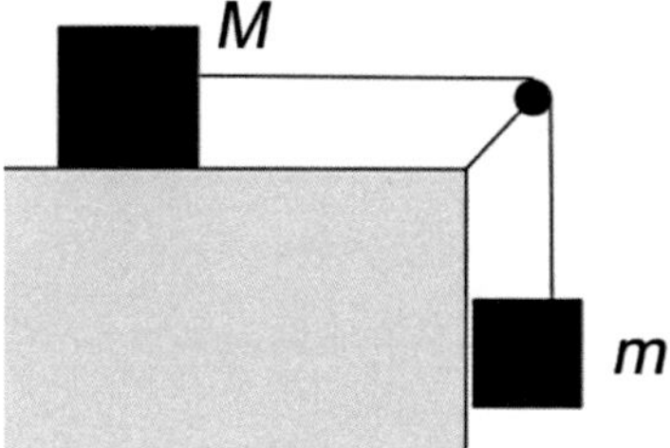

Solution:
Force acting on mass M: $T = Ma$. Force acting on mass m: $T - mg = ma$. Therefore,

$$Ma - mg = ma \quad \text{or} \quad a(M - m) = mg \quad \text{and}$$

$$a = mg/(M - m).$$

CONCEPT CHECK
1. **What** is the difference between Newton's first and second laws?
2. **Why** is mass more fundamental than weight?

2.3 Newton's Third Law

LEARNING OBJECTIVES
1. **Describe** Newton's third law.
2. **Explain** how rockets and airplanes obtain their thrust force.

So far we have focused on single forces, but nature is full of symmetries. Newton recognized that it is impossible to have only a single force. When you push on a wall, the wall pushes back. Anytime a force acts, a reaction force acts as well. Newton saw this symmetry. Examples ranging from walking to jet and rocket propulsion remind us daily of this simple, yet powerful principle.

Action and Reaction Force Pairs

In nature, *forces always exist in pairs*. We can describe these forces as action and reaction pairs. Newton described these force pairs in his third law of motion. Note, however, that these pairs do not cancel each other out: action and reaction forces are equal and opposite in nature, but they act on different objects.

Newton's third law of motion If an object exerts a force on a second object, the second object exerts an equal and opposite force back on the first object.
The action-reaction terminology is somewhat misleading: it implies that one force comes into being in response to the other. But in fact, both forces come into existence simultaneously: naming one the action and the other the reaction is really arbitrary.

free-body force diagram A diagram showing all forces acting on a single object.

We can improve the clarity and our understanding of many physical problems by drawing forces on each object of the problem separately. We call such drawings **free-body force diagrams**. See **Example 2.3.**

Example 2.3
A tow truck pulling a car
A tow truck with a mass of 3000 kg is pulling a 1500-kg car. Both vehicles accelerate at the rate of 5 m/s^2. What is the force in the cable that pulls the car? What is the driving force of the truck?

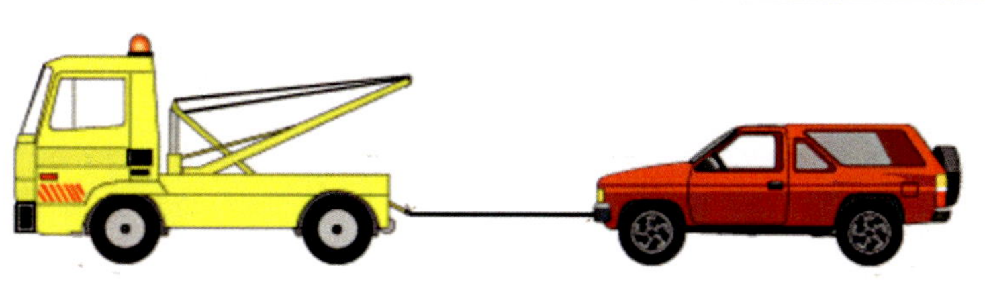

Solution:

The force in the cable is also known as tension. Tension is a pulling force in a string, cable, or belt. It acts along the cable and disappears unless the cable is taut.

Applying Newton's third law to the force of the cable on the truck and the car, we see that the truck is pulling on the car with the force T, and the car is pulling back on the truck with an equal and opposite force.

Let us draw a free-body force diagram object separately. The truck has two forces acting on it.

Now we apply Newton's second law to the motion of the truck:

Driving force of the truck – force in the cable = mass of the truck × acceleration of the truck

In symbols,

$$F - T = Ma$$

Only one force acts on the car.

Now we apply Newton's second law to the motion of the car:

Force in the cable = mass of the car × acceleration of the car

In symbols,

$$T = ma = (1500 \text{ kg})(5 \text{ m/s}^2) = 7500 \text{ N}$$

We can go back to the first equation describing motion of the truck and substitute for the variables that we know:

$$F - T = Ma = F - 7500 \text{ N} = (3000 \text{ kg})\ (5 \text{ m/s}^2) = 15000 \text{ N}$$

And therefore:

$$F = 22500 \text{ N}.$$

We conclude that the force in the cable is 7500 N and the driving force of the truck is 22500 N.

Rockets and Airplanes

Imagine you are stranded on a frozen lake. If you try to walk, your feet just slip on the ice. How can you get back to the shore? You could try this: take off your shoes and throw them in the direction that is opposite the way you want to go. As you throw each shoe, you exert force upon it. At the same time, the shoe exerts a force back on you, and you accelerate toward the shore – in your socks! You have just experienced the principle behind a rocket thrust.

Rockets take along their own material to eject. Rocket fuel is hydrogen and oxygen stored in liquefied form at very low temperatures. When combined, their combustion produces steam, which blasts out of the back of the engine. The velocity of the exhaust can reach over 16,000 km/h (about 10,000 mph). This creates an enormous push in the opposite direction: a thrust. This action-reaction, familiar to us from our study of Newton's third law, is similar to what happens if we blow up a balloon and release it. See **Figure 2.5**.

The principle of forward propulsion in airplanes is slightly different. A propeller-driven airplane uses its propeller to push air back at a higher speed. As the propeller pushes back on the air, the air in turn pushes forward on the propeller and therefore the plane. This force provides the forward thrust. The jet plane is an improvement on the propeller-driven airplane, but it uses a similar principle. Both propeller- and jet engine-driven airplanes need atmosphere. Our atmosphere gets thin once you get above an altitude of 10 km, so these planes could not travel in space.

Balloon and rocket thrust Figure 2.5

The action-reaction force pairs act simultaneously. They are numbered here to assist clarity.

2. Reaction: The rushing air exerts an equal and opposite upward force on the balloon. Balloon flies up.

1. Action: Balloon exerts downward force on the air as it escapes the balloon, rushing downward.

2. Gas pushes back on the rocket with an equal and opposite force, generating the rocket thrust.

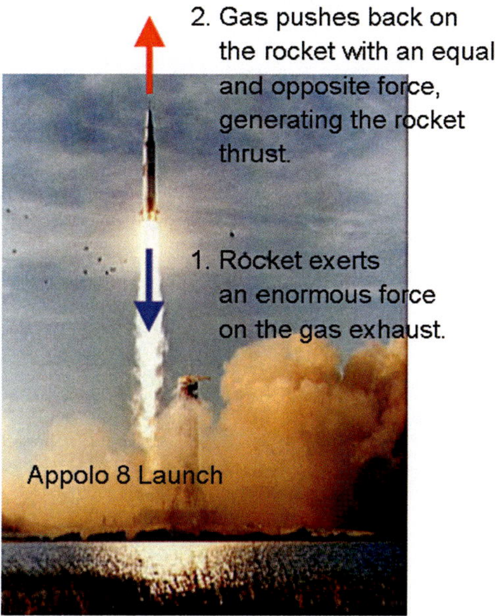

CONCEPT CHECK

1. **How** can you apply Newton's third law to explain the driving force action of a car?
2. **What** is the principal difference between the propulsion of a jet plane and a rocket?

2.4 Friction and Air Drag

LEARNING OBJECTIVES

1. **Describe** static and sliding friction.
2. **Explain** terminal speed.

When surfaces slide over one another, an opposing force of friction is present. We often think of friction as a handicap: engineers seek to avoid wear and tear in machines by designing them with a minimum of friction between moving parts; machine operators use lubricants to make mechanical parts run more smoothly and efficiently. But without friction, we could not do many of the things we take for granted. We could not walk because we could not push backward on the floor and the floor could not push back on us. Car and bicycle wheels would slide rather than roll. Our cars could not brake. Nails hammered into walls would not stay in place without friction. Rock climbing would be almost impossible.

Static and Sliding Friction

We already know that friction is a force that usually opposes motion between sliding objects. But an object does not need to slide to experience friction. Friction can also be a reaction force that opposes a slight push on stationary object, preventing the object from moving. Irregularities in the surfaces in contact with one another are the primary cause of friction (**Figure 2.6**).

Microscopic irregularities cause friction Figure 2.6

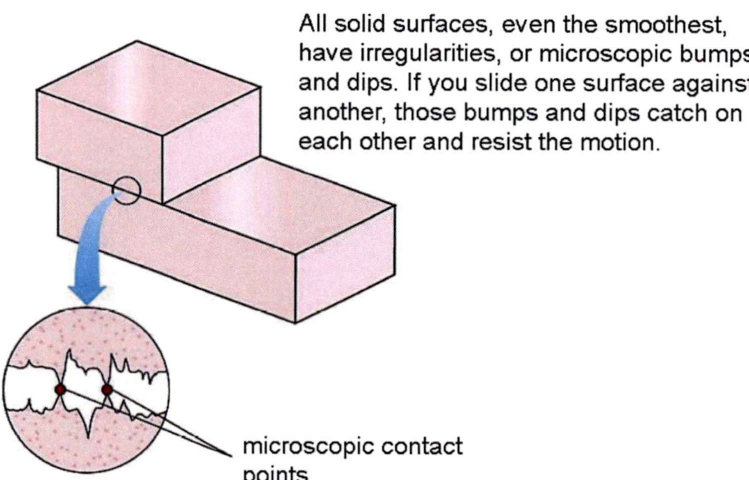

Let us do a simple experiment to understand two types of friction. First place a textbook flat on a desk and push it very gently. Nothing happens: an opposing force of **static friction** between the book and the desk counterbalances your force. The net force on the book is zero. Now push harder, and the textbook starts sliding. You have just overcome the maximum opposing static force of friction.

static friction A reaction friction force to a push on a stationary object; friction force between two solid objects that are not sliding relative to each other.

As you continue to push on the sliding book, it feels as if you no longer have to push so hard to keep the book sliding. This is even more noticeable if you push on something bigger, like a large piece of furniture. The frictional force now changes to **sliding friction.** Sliding friction is a little smaller than the maximum static frictional force that you had to overcome to get the textbook moving. See **Figure 2.7**.

Static and sliding friction acting on a book Figure 2.7

Apply gentle force to a book lying flat on a table. The book does not move: static friction equals the force on the book.

force on the book

static friction

friction force

maximum static friction

static friction = applied force

static friction

applied force on book

no motion

book sliding

Increase the applied force. The book still does not move: static friction again equals the force on the book.

static friction

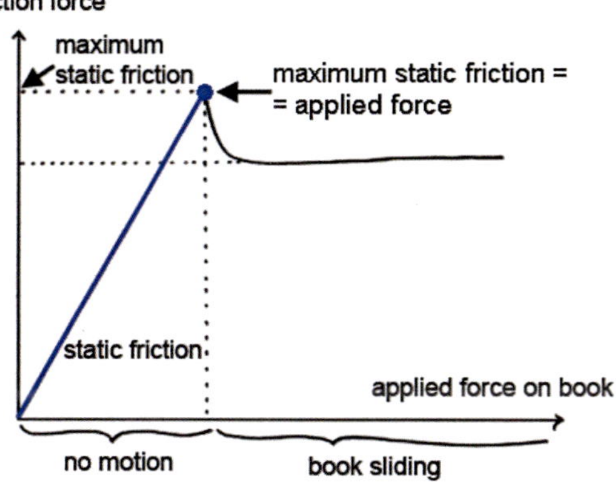

Now slightly increase the applied force on the book. The book begins to slide. A force of sliding friction still opposes the book's motion, but it is smaller than the maximum force of static friction.

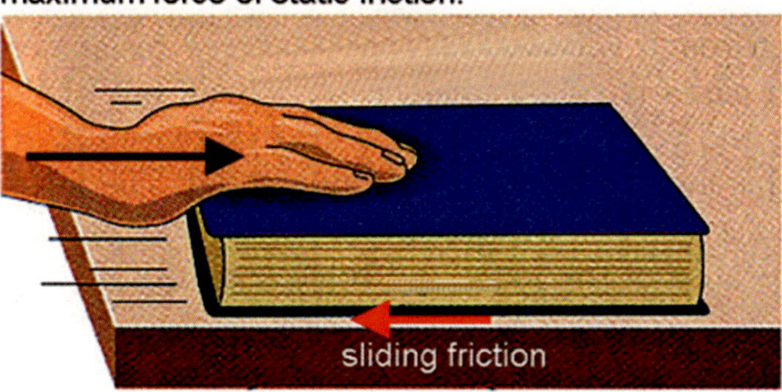

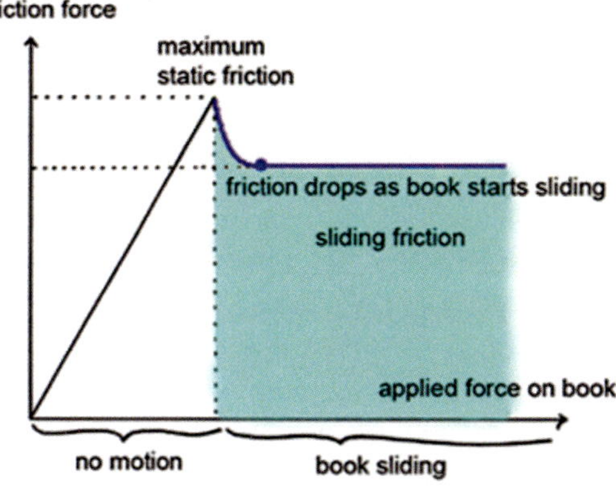

This sliding frictional force is present whenever an object slides along any surface that is not perfectly smooth.

sliding friction Force exerted by any surface opposing a sliding motion.

How will the friction change if we push down on the book? Pressing any two surfaces together harder as you slide one surface against the other will cause the resistance to increase. The number of contact points increases, which in turn increases the friction. By pushing down on the book, you increase the reaction force of the surface pushing back on the book. The reaction force is also called a **normal force**, and it is perpendicular to the surface. Both static and sliding friction increase in direct proportion to the normal force. For instance, by doubling the normal force when we press on the book, we also double the force of static or sliding friction (**Figure 2.8**).

normal force The force of a surface acting on an object and perpendicular to the plane of contact.

Normal force and friction Figure 2.8

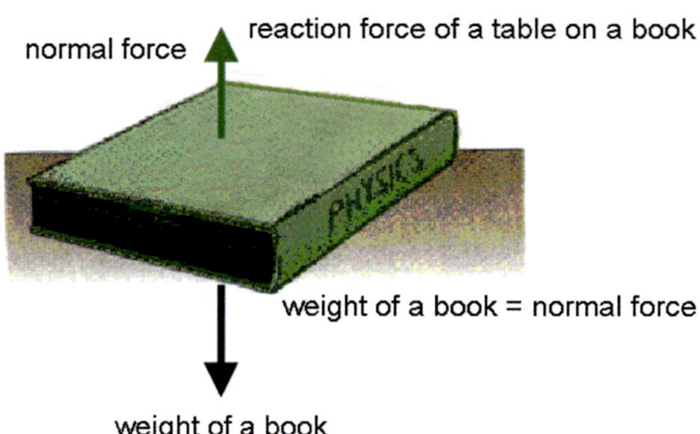

You can increase the normal force acting on the book by pushing down on the book.

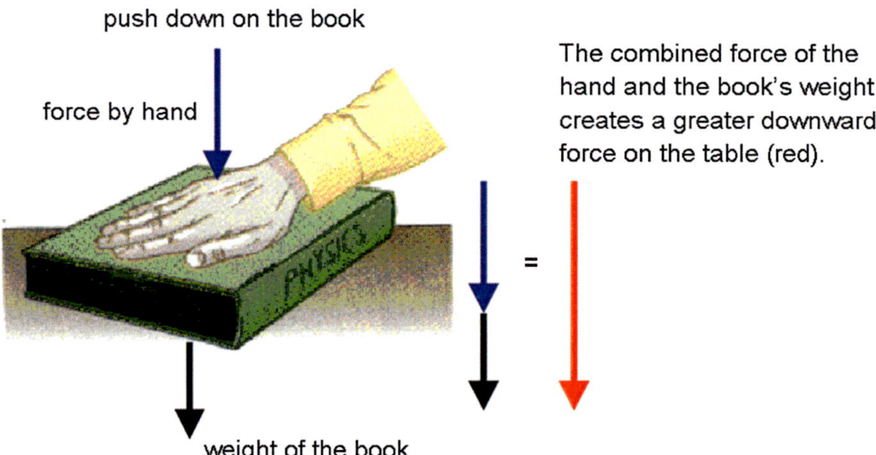

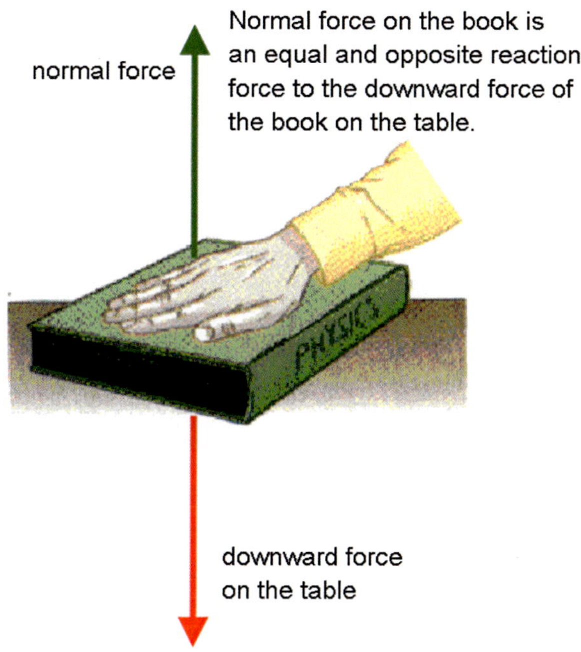

Both static and sliding friction on the book increase in direct proportion to the increase of the normal force.

So far we have learned that frictional force depends on how hard an object is pushed onto its supporting surface. Friction also depends on the smoothness of the surface and is proportional to the **coefficient of friction,** or **μ,** for the surface. We describe the frictional force f as

$$\text{Friction} = \text{coefficient of friction} \times \text{normal force}$$

In symbols,

$$f = \mu N$$

The coefficient of friction is dimensionless and is determined by the smoothness of the surface. It has two forms: static and sliding. For instance, the coefficient of static friction between your shoes and ice is 0.1, and the coefficient of sliding friction is 0.05. Similarly, car tires on dry concrete have static and sliding coefficients of 1.0 and 0.8, respectively. (See **Table 2.1**.) Speed skaters look for ways to minimize sliding friction and like to skate on ice whose temperature is slightly above the melting point. A film of moisture forms on the ice's surface, and the skaters' blades glide better on this wet ice than on the dry ice that would occur in subfreezing temperatures.

Static friction can take on a range of values but must be always less or equal to the static coefficient times the normal force or

$$f_s \leq \mu_s N.$$

On the other hand, the kinetic or sliding friction can take on only a singular value given by

$$f_k = \mu_k N.$$

Coefficients of static and sliding friction Table 2.1

	Static μ_s	Sliding μ_k
average tire on dry pavement	1.0	0.8
grooved tire on wet pavement	0.8	0.7
glass on glass	0.9	0.4
metal on metal (dry)	0.6	0.4
smooth tire on wet pavement	0.5	0.4
metal on metal (lubricated)	0.1	0.05
steel on ice	0.1	0.05
shoes on ice	0.1	0.05
steel on Teflon	0.05	0.05

It is interesting that the force of friction does not depend on the surface contact area or speed. A crate or a car skidding at low speed has approximately the same friction as the same crate or car skidding at high speed. If you slide a crate on its smallest surface, you concentrate the same weight on a smaller area with the result that the friction is the same (**Figure 2.9**). **Example 2.4** shows a simple application of Newton's second law that involves the force of sliding friction.

Friction is independent of surface area for hard contact surfaces Figure 2.9

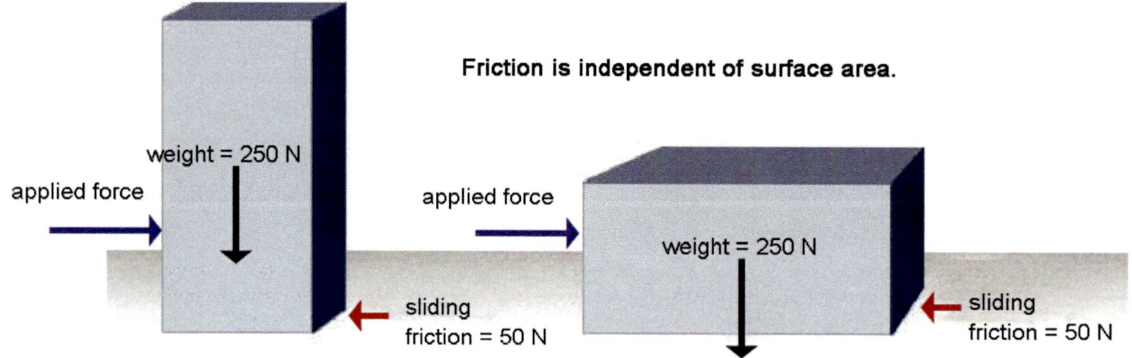

Example 2.4
Friction acting on a sliding crate

A man is pulling on a 100-kg crate that slides along the floor with a constant velocity. If the sliding coefficient of friction is 0.4, what is the sliding frictional force opposing the motion, and how much force does the man need to apply?

Solution:
The free-body force diagram shows that
Net force on the crate = applied force - friction = mass × acceleration.

Since the velocity of the crate is constant, the acceleration = 0, and
Applied force = friction = coefficient of friction × normal force.

The reaction or normal force of the floor on the crate equals its weight, and weight = mass × acceleration due to gravity = mg = (100 kg)(9.8 m/s^2) = 980 N.

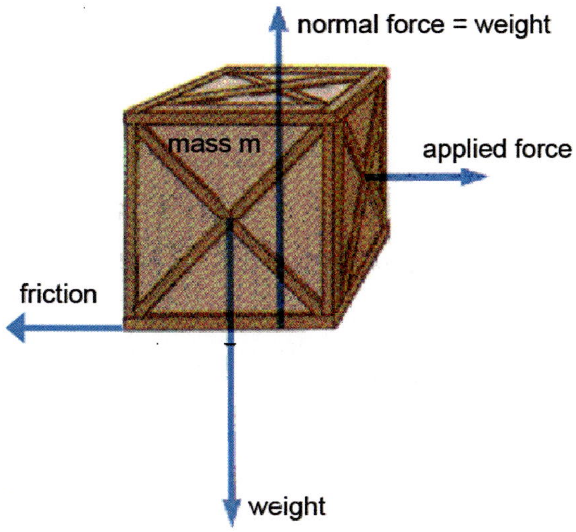

Therefore, normal force = 980 N and sliding friction = $\mu_{sliding} \times$ N = (0.4)(980 N) = 392 N

The applied force must be 392 N also.

We can apply the concept of friction to an object sliding down an inclined plane. See **Example 2.4**. In **Example 2.5**, we show how friction stabilizes a ladder propped against a wall.

Example 2.4
Inclined plane
An object of mass *m* is sliding down an inclined plane, whose angle of inclination is θ and the coefficient of sliding friction is μ. Find its acceleration *a*.

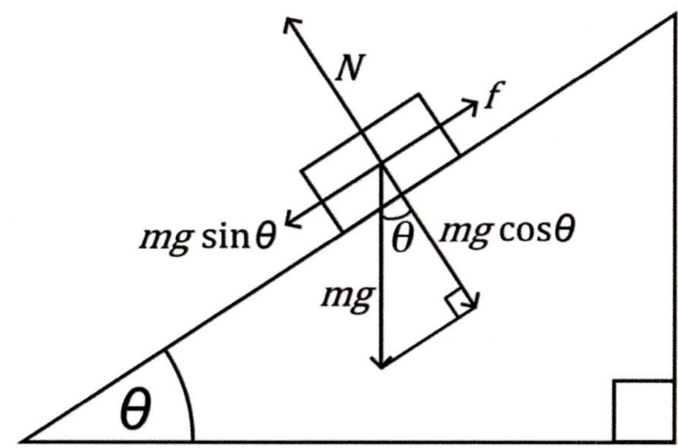

Solution:

$$F = mg\ sin\theta - f = ma \quad or \quad mg\ sin\theta - \mu N = ma.$$

But

$$N = mg\ cos\theta.$$

Therefore,

$$mg\ sin\theta - \mu mg\ cos\theta = ma.$$

And

$$a = g \sin\theta - \mu g \cos\theta.$$

Example 2.5
Friction acting on a ladder propped against a wall
Identify all forces acting on a ladder propped against a wall.

Solution:
Both ends of the ladder experience a normal force of the floor and the wall acting perpendicularly on the ladder. If both the floor and the wall are not perfectly smooth, both surfaces provide friction that opposes the sliding motion of the ladder ends while in contact with the floor and the wall.

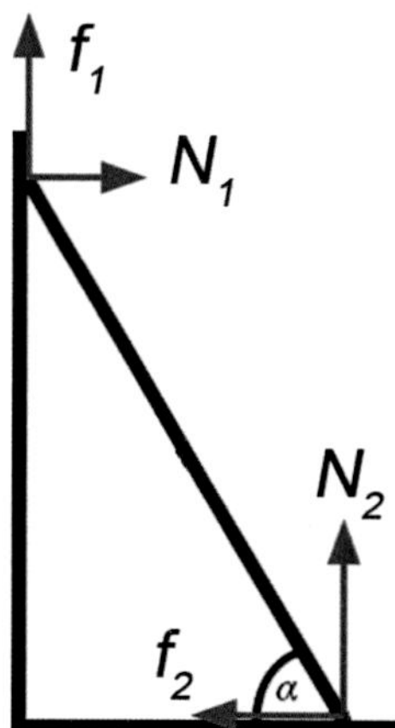

One might think that if the sum of all forces acting on the ladder is zero, then the ladder will be in equilibrium and will not crash. That may or may not be true because it is not the sufficient condition. In the Rotation chapter we will learn another condition that will guarantee the ladder's static equilibrium.

We have studied how friction acts on static and sliding objects, but how about rolling objects? Without friction, wheels and round bodies would slide, not roll. The frictional force between any car's tires and the road must be static. Otherwise the car would start sliding and the driver would lose control.

A rolling motion has one advantage over sliding. Rolling friction is lower than sliding friction. Therefore replacing a sliding motion with rolling reduces friction between moving mechanical parts and saves energy. The Ancient Egyptians knew this trick for reducing friction: when building the pyramids, they moved huge stone blocks on rolling logs. We use ball bearings or roller bearings, in many mechanical gadgets to reduce friction.

Air Drag and Terminal Speed

Friction is not restricted to solid surfaces sliding or rolling over one another. Friction occurs also in liquids and gases. If you have traveled in a car and stuck your hand out the window, you have felt air resistance from the fast, oncoming air hitting your hand. Air resistance comes from collisions with air molecules. We know that in a vacuum, all bodies accelerate downward at the same rate. On the surface of the Earth and just above, that rate is 9.8 m/s^2. However, when an object falls through the air, it experiences a retarding force of air resistance (**Figure 2.10**).

Terminal speed in skydiving Figure 2.10

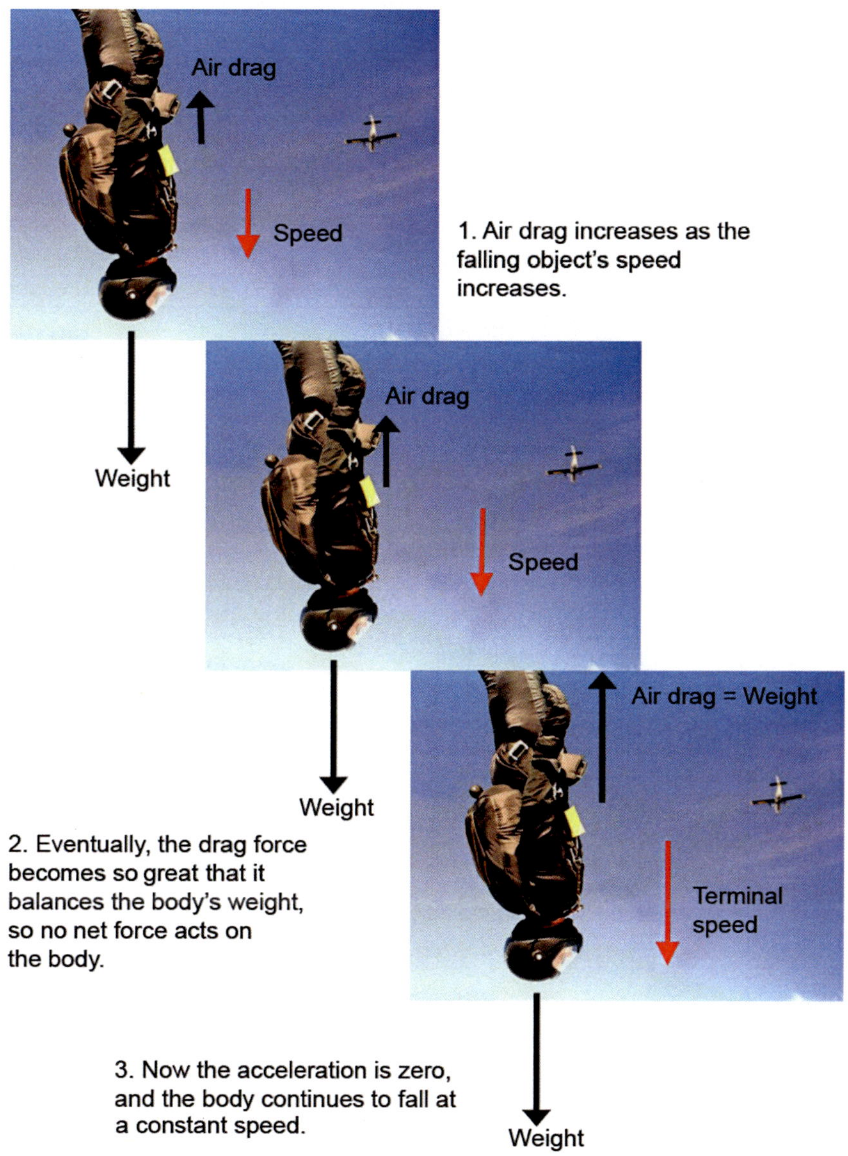

This force increases as the object's speed increases. Eventually the force becomes so great that it balances the force of gravity. The object's acceleration is reduced to zero, and it continues to move at a constant speed. This speed is called the **terminal speed**.

terminal speed The final speed a falling object reaches when the air drag force exactly opposes the weight of the object.

We already know that a higher speed leads to a higher air drag. A larger frontal area or denser air contributes to greater air resistance as well. The projected frontal area of the falling object A, the speed of the object v, and air density ρ—all combine to determine the total **air drag** F_D, which is given by

$$F_D = C_D A \rho v^2/2.$$

Here C_D a drag coefficient. More aerodynamic objects have lower drag coefficients and therefore present lower air drag.

CONCEPT CHECK

1. **What** type of friction between tires and the road enables us to ride bicycles?
2. **How** could you maximize your terminal speed if you were a speedskier?

2.5 Centripetal Force

LEARNING OBJECTIVES

1. **Explain** the centripetal force.
2. **Describe** the force needed to maintain circular motion.

The motion of the Earth around the Sun, the Moon or a satellite around the Earth, a race car around a circular track, or a twirling ball attached to a string above your head are all examples of circular (or nearly circular) motion. They occur in a two-dimensional plane.

Newton's first law tells us that at a constant speed, straight-line motion implies the absence of a net force. But while in circular motion, an object is continuously changing direction. It must have a net force acting on it at all times. Imagine a ball that you twirl on a string above your head. The string is exerting force on the ball, and this force can only exist along the string's length. The force must act toward the inside of the turn to keep the ball turning toward you. It acts toward the center of the circle and we call it the **centripetal force** (**Figure 2.11**).

centripetal force Center-seeking force needed to maintain circular motion

Circular motion requires centripetal (center-seeking) force. Figure 2.11

If you cut the string, inertia will make the ball fly off in a straight line and continue traveling along the direction of velocity the ball had when the string was cut.

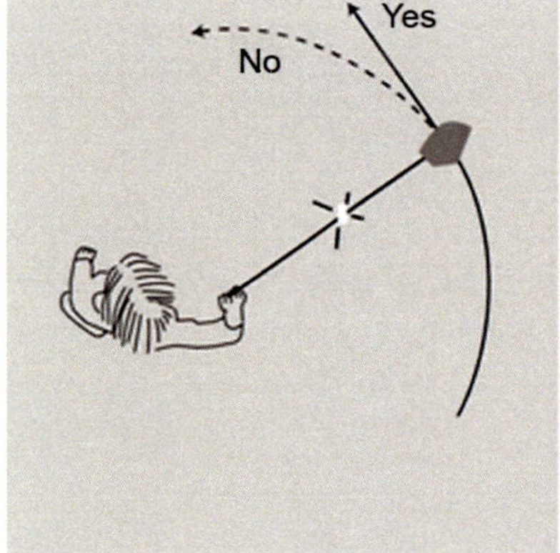

Centripetal force is needed for circular motion.

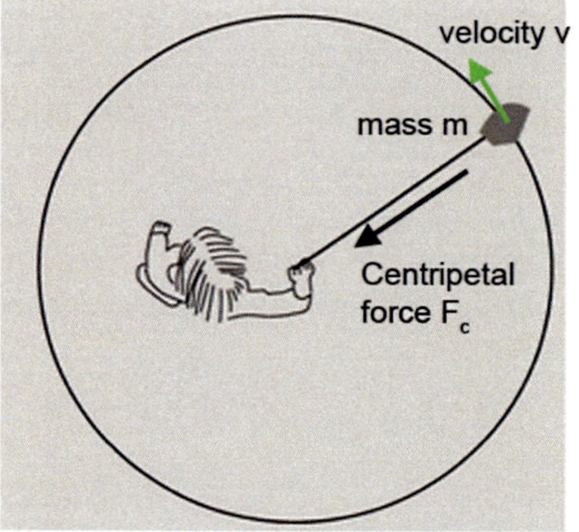

Force directed toward the center of the turn must act to keep changing the object's velocity direction and provide acceleration directed toward the center of the turn as well.

There are other examples of centripetal force. Centripetal force helps separate water from your wet clothes in your washer. During the spin cycle of the washer the tub spins at high speeds and exerts a large centripetal force on the clothes. But the holes in the tub prevent the application of the same force on the water in the clothes. The water escapes tangentially through the holes.

Before we look at more examples of circular motion, we can develop a deeper understanding of it by considering what variables determine the magnitude of the centripetal force. Let us perform a simple experiment. For a given length of the string twirl the ball at different speeds. You will notice that as you twirl it faster you have to exert a larger force on the string. Now shorten the length of the string. You will feel that it is a little harder to twirl, and you will again need to exert a larger force. This simple experiment leads you to the conclusion that increasing the **tangential speed** or decreasing the radius of rotation increases the required centripetal force.

tangential speed The linear speed tangent to a curved path, such as a circle.

The centripetal force F needed to keep an object of mass m moving in a circle of radius r equals mass times tangential speed of the object v squared, divided by the radius r or

$$\text{Centripetal force} = \text{mass} \times \frac{(\text{speed})^2}{\text{radius}} \quad \text{or} \quad F_c = m\frac{v^2}{r}$$

Changes in speed have the greatest impact on the centripetal force needed to maintain circular motion. Doubling the speed increases the required centripetal force by a factor of four, while doubling the radius reduces the centripetal force by a factor of only two.

It is the frictional force between the tires and the road that provides the turning or centripetal force. To round a turn safely, we can choose to slow down. That way we will need a smaller centripetal force to make the turn, and therefore less friction between the tires and the road. Another option is to round a wider turn. This would increase the radius and thus decrease the centripetal force needed as well. A banked turn will help the car round the turn faster. Many car racetracks have banked turns to allow racecars to move faster (**Figure 2.12**).

Banked turn forces acting on a car. Figure 2.12

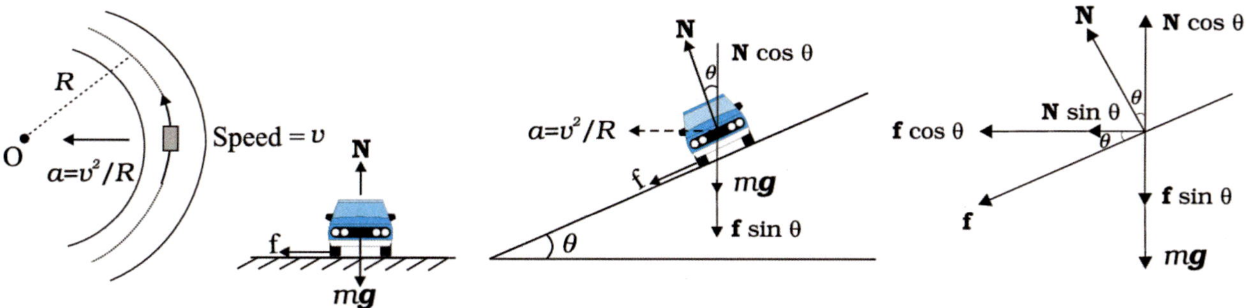

The normal force of the road acting up on the car and the car's weight do not act along the same vertical line as is the case for a flat turn. The normal force has a component pointing toward the center of the turn. This provides an extra center-seeking force in addition to the friction between the road and the tires and the car can go into the turn faster.

If you are on a fast-moving, spinning carnival ride, you may feel as if there were some kind of outward force acting on you. The ride pulls you in toward the center, but you get the sensation that there is a force that is trying to push you outward. Physicists refer to this apparent force as a centrifugal force. But this is an unfortunate choice of a name because what you experience is not a true force. It is simply a consequence of Newton's first law. At any point, you have a tendency to continue moving along a straight line, tangential to the circle, instead of turning to follow the circular trajectory.

CONCEPT CHECK

1. **What** kind of force is needed to maintain circular motion?
2. **How** does a banked turn help a car round a turn faster?

Summary

2.1 Newton's First Law

- Every object continues in a state of rest or in a state of motion at a constant velocity (constant speed in a constant direction) unless acted on by a net force.
- **Inertia** is the natural tendency of an object to remain at rest or in motion at a constant velocity. The **mass** of an object is a quantitative measure of inertia.

2.2 Newton's Second Law

- **A n**et force acting on an object produces an acceleration that is directly proportional to the magnitude of the force and in the same direction as the force. The greater the net force, the greater the acceleration produced.
- The acceleration produced by a net force acting on an object is inversely proportional to the mass of the object. The greater the mass of the object, the smaller the acceleration produced.

$$\text{acceleration} = \frac{\text{net force}}{\text{mass}} \qquad \text{or} \qquad \text{net force} = \text{mass} \times \text{acceleration}$$

In symbols:

$$a = \frac{F_{net}}{m} \text{ or } F_{net} = ma$$

- **Weight**

$$\text{Weight} = \text{mass} \times \text{acceleration due to gravity}$$

In symbols,

$$w = mg$$

- While mass will always stay the same, weight will change with the size of the gravitational acceleration.

2.3 Newton's Third Law

- If an object exerts a force on a second object, the second object exerts an equal and opposite force back on the first object.

2.4 Friction and Air Drag

- Pushing on a static object causes an opposing **static friction force**. **Sliding friction** slows down a sliding object if the surfaces are not perfectly smooth.
  We describe the force of friction f as

  $$\text{Friction} = \text{coefficient of friction} \times \text{normal force}$$

  In symbols, $$f = \mu N$$

  where μ is coefficient of friction and N is the **normal force.**
- An opposing air drag force acts on objects moving through the air. When the air drag force matches the weight of a falling object, the falling object reaches its **terminal speed.**

2.5 Centripetal force

- The **centripetal force** F needed to keep an object whose mass is m moving in a circle with radius r equals mass m times **tangential speed** of the object v squared, divided by the radius r.

$$\text{Centripetal force} = \text{mass} \times \frac{(\text{speed})^2}{\text{radius}} \quad \text{or} \quad F_c = m\frac{v^2}{r}$$

Key Terms

force
Newton's first law
net force
inertia
mass
Newton's second law
free-body force diagram
weight
Newton's third law
static friction
sliding friction
normal force
terminal speed
centripetal force
tangential speed
drag force

Critical Thinking Questions

1. How can you tell when a body is exerting a net force on another body?
2. A car is standing parked on a street when another car rear-ends it. Describe what happens to both drivers and why.
3. A crate on the floor has the force of its own weight acting on it. Why is it not moving, as Newton's first law might suggest?
4. Why does your body lean to the outside of the curve when your car makes a sharp turn?
5. You slide a box along the floor, and then you flip it so that a different side is in contact with the floor. Will the frictional force change?
6. A crate that is heavier than you are rests on a rough floor. The coefficient of static friction between the crate and the floor is the same as that between the soles of your shoes and the floor. Can you push the crate across the floor?
7. If two objects accelerate at the same rate, must they have the same mass? Must they have the same net force acting on them?
8. Would perfectly frictionless shoelaces stay tied?
9. You shake a beach towel to get rid of sand. What principle of physics are you applying?

10. Describe what happens when you speed into a turn where oil has spilled on the pavement.
11. As a skydiver, how would you position your body in the air to reach the largest possible terminal speed?
12. When a rocket burns its fuel, its mass decreases; how does this affect the rocket's acceleration?
13. If a couple of sumo wrestlers ride in a compact car, what happens to its acceleration? Assume the driving force does not change.
14. Explain how a person trapped in the middle of a frictionless ice rink can get to the side of the rink.
15. How could you decide which of two masses is greater if you could apply a constant force to each?
16. Why does a banked track support higher speeds in the turns?
17. What causes people on a wave-rider to feel an outward "force"?

Exercises

1. You are standing on a frictionless skateboard. You push against a wall with a force of 35 N. If your mass is 65 kg, what is your acceleration?
2. A bicycle racer accelerates at the finish line with a force of 100 N. A wind opposes his motion with a force of 20 N. The racer's mass is 65 kg. What is his acceleration?
3. You are standing on ice. Two of your friends are pushing you with two different forces. One is pushing westward with a force of 50 N, and the other is pushing northward with a force of 50 N. What are the direction and magnitude of the resultant force?
4. A car pulling a trailer has a total mass of 2500 kg and can accelerate from rest to 19 m/s in 9 seconds. What is the net force on the car?
5. A tow truck with a mass of 2500 kg is pulling a 1500-kg car with a force of 20,000 N. What is the acceleration of the car?
6. The strength of gravity on Mars is 40% of that on Earth. If a girl has a mass of 30 kg, what would be her weight on Mars?
7. A skydiver has a mass of 70 kg and a net acceleration of 7 m/s^2. How great is the air resistance she experiences?
8. What is the static frictional force acting on you when you stand on ice if your mass is 70 kg and the coefficient of static friction is 0.1?
9. What is the sliding frictional force that would act on a 50-kg moving crate if the coefficient of sliding friction were 0.15?
10. The above crate has a maximum static friction of 100 N. You push on it with a force of 110 N. What is the crate's acceleration?
11. A dog sled on snow has a static friction coefficient of 0.15. The loaded sled has a total mass of 39 kg. Determine the horizontal force needed to budge the sled
12. When a parachute opens, it experiences a large upward drag force. This upward force is initially greater than the weight of the skydiver, and it briefly jerks him upward. If the weight of the skydiver is 900 N and the opposing drag force is 1000 N, what are the magnitude and direction of the acceleration?
13. What is the centripetal force acting on an 80-kg person on a wave-rider sitting on the perimeter of a wheel 15 m in diameter that is moving with a tangential speed of 8 m/s?

Chapter 3
Gravity

One of the early applications of Newton's laws of motion was his explanation of the Moon's circular motion around Earth. Newton proposed a law of universal gravitation that explains not only the behavior of objects falling near the surface of Earth, like an apple falling from a tree, but also why the Moon orbits Earth and why planets move in curved paths about the Sun. Today, the law of gravity provides an important foundation for determining orbits of satellites around the Earth, orbits of all planets in our solar system, or movement of any matter in the Universe.

3.1 Gravitational Force

LEARNING OBJECTIVES

1. **Describe** a gravitational force.

Sir Isaac Newton discovered the central idea of gravity during his stay on his family's farm in the late 1600s. According to his recollections, the idea came to him one night when he was sitting under an apple tree: as he was watching the Moon, an apple fell and barely missed his head. At that point, Newton realized a similarity between the apple and the Moon. The apple was falling toward the ground, accelerated by the Earth's gravitational pull, and so was the Moon.

Law of Universal Gravitation

The connection between the falling apple and the motion of the Moon is not easy to see because the Moon is orbiting the Earth. The direction of its velocity is around the Earth rather than toward its center. It keeps moving in a circular orbit just like a ball attached to a rope and twirled above your head. The force of the tension in the rope keeps the ball moving in a circle and is directed toward the center of the circle. Newton realized that there must be a similar force acting on the Moon as well, and this force must be directed toward the center of the Earth, just as it is in the case of the falling apple. The gravitational force acts on the Moon in the same way that it acts on the apple. But the Moon does not crash into the Earth because its tangential speed around the Earth is so large that the distance it falls toward the Earth each second is the same as the distance the Earth's surface curves away from the Moon. Thus, the falling Moon never catches up to the Earth. To arrive at that conclusion, Newton conducted a simple thought experiment of a cannon ball launched at increasing speeds from the top of a tall mountain. See **Figure 3.1**.

The idea that the force that shapes the Moon's path is the same force of gravity that pulls the apple to the ground was a brilliant insight. It was difficult to believe that this same force acting on the apple would also act on the Moon more than 240,000 miles away. Newton reasoned that the force of gravity between any two objects was proportional to the product of their masses. But how would this force scale with distance? Newton knew that the distance to the Moon is about 60 times larger than the Earth's radius. He also suspected that the force of gravity is inversely proportional to the squared distance between the two objects. If this were true, then the acceleration of any object 60 Earth radii away would be 3600 (60^2) times smaller than that on the surface of the Earth, which is 9.8 m/s^2. Thus Newton expected to get 9.8/3,600 = 0.0027 m/s^2 for the acceleration of any object 60 Earth radii away from the Earth, such as the Moon. To check his argument, he had to find the acceleration of the Moon. **Example 3.1** shows how he did it.

Newton's thought experiment Figure 3.1

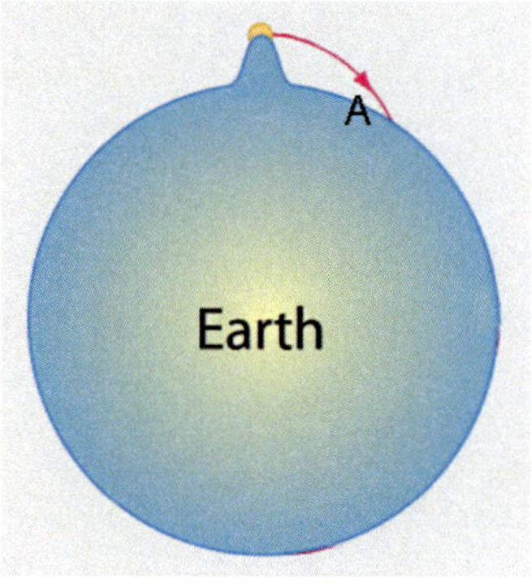

1. A cannon ball is shot and travels a certain distance (path A).

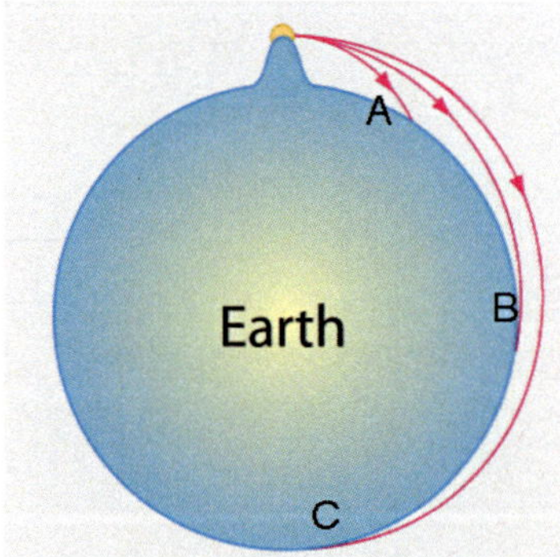

2. If more gunpowder were used, the cannon ball would travel with greater speed and go even farther (paths B and C).

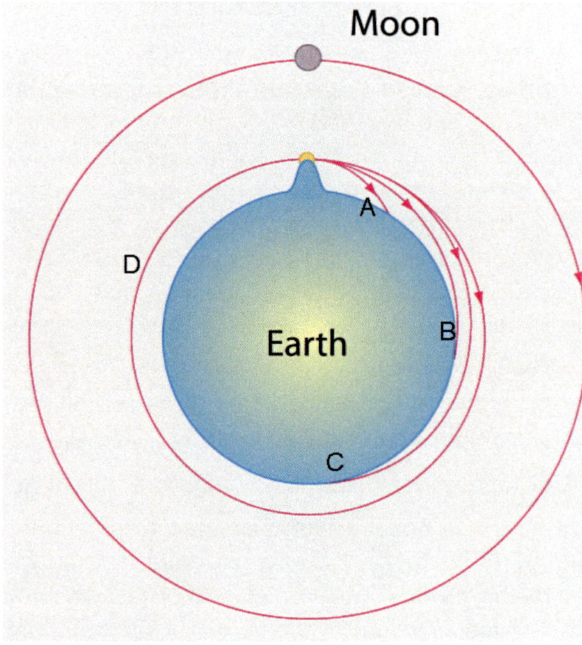

3. If the cannon ball were fired with even greater force and traveled with greater speed,it might curve around Earth and hit the back of the cannon (path D). It would be in a circular orbit. The ball would never come down and would continually miss hitting Earth – just like the Moon.

Critical thinking question...

What would happen to the shape of the orbit if we increased the speed further, beyond that used to get into orbit D?

Example 3.1
Acceleration of the Moon
If the Earth-Moon distance is 3.84×10^8 m, and it takes the Moon 27.3 days to go around the Earth, find the acceleration of the Moon. While Newton did not know these numbers precisely, he knew them closely enough to undertake this calculation.

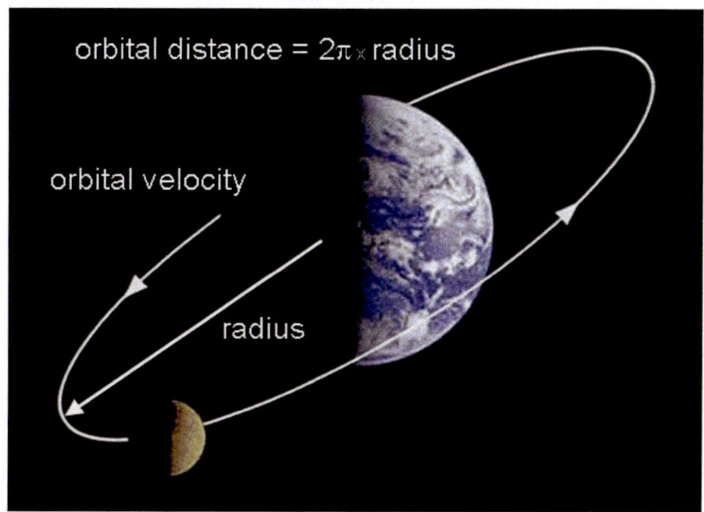

Acceleration for any circular orbit is derived as velocity squared divided by the radius of the orbit. You can find the length of the Moon's orbit by noting that the circumference of a circle is given by $2\pi \times$ radius of the circle. If the Earth-Moon distance is 3.84×10^8 m, and it takes the Moon 27.3 days to go around the Earth, we can find the Moon's velocity. But first we need to convert 27.3 days into seconds.

$$27.3 \text{ days} = 27.3 \text{ days} \times \frac{24 \text{ hours}}{\text{day}} \times \frac{3600 \text{ seconds}}{\text{hour}} \simeq 2.36 \times 10^6 \text{ seconds}$$

$$\text{Orbital velocity of the Moon} = \frac{\text{distance traveled}}{\text{time}} = \frac{2\pi(3.84 \times 10^8 \text{meters})}{2.36 \times 10^6 \text{seconds}} = 1022 \text{ m/s}$$

Now you can use this magnitude of velocity to find the Moon's acceleration in its circular orbit.

$$\text{Acceleration} = \frac{(\text{orbital velocity})^2}{\text{radius}} = \frac{(1022 \text{ m/s})^2}{3.84 \times 10^8 \text{ m}} = 0.0027 \text{ m/s}^2$$

This simple calculation confirmed Newton's hunch that the gravitational force decreases in inverse proportion to the distance squared.

Newton's thought experiment (**Figure 3.1**) showed that a projectile shot at a high enough speed can reach a circular orbit around the Earth. **What a Physicist Sees** examines the circular orbit around the Earth of a space shuttle and shows how astronauts inside experience **weightlessness**.

weightlessness Being without a support force, as in free fall.

What a Physicist Sees
Circular orbit around the Earth and weightlessness
The Space Shuttle orbits roughly 150 miles above the Earth. The astronauts inside the orbiting Space Shuttle or the International Space Station experience artificial weightlessness even though the astronauts are only 150 miles above the surface (Figure a).

a.

The Space Shuttle with the astronauts inside is in a free fall, just as you would be in a falling elevator. But it never gets closer to the Earth because its tangential velocity is so large that the distance it falls within 1 second is the same as the distance the Earth curves away due to its spherical shape. See **Figure b**.

b.

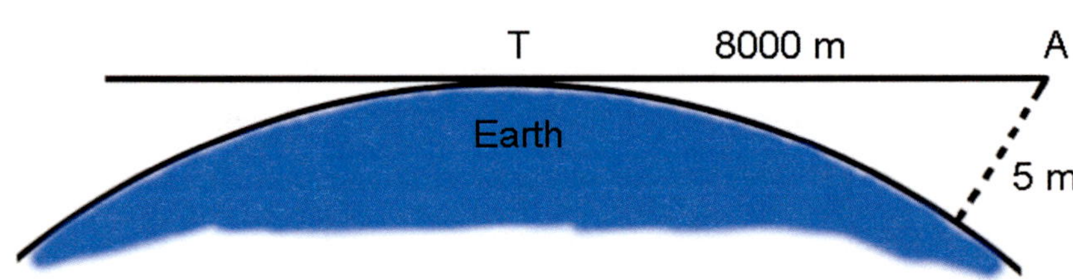

If you draw a line tangent to the Earth's surface (line TA), the surface drops a vertical distance of 5 meters for every 8000 meters tangent to the surface.

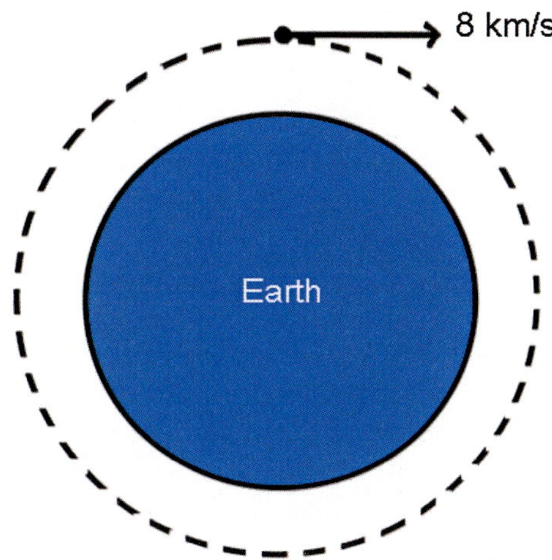

The space shuttle has a tangential speed of 8 kilometers per second, during which time it also falls 5 meters because the vertical distance of its drop $d = \frac{1}{2} a t^2 \cong 5 t^2$ and $t = 1$ second. The shuttle never catches up with the surface of the Earth even though it is in a continuous free fall.

Think Critically...
The Earth is pulled in toward the Sun by a strong gravitational force. Why doesn't it crash into the Sun?

Having made the connection between the motion of the Moon in the heavens and the motion of an apple near the Earth's surface on the one hand and a force that reaches across an empty space and pulls both to the Earth on the other, Newton made another step. He stated that the force of gravity exists between all objects everywhere in the Universe. In 1687 Newton published his findings in his book *Philosophiae Naturalis Principia Mathematica*, where he presented not only the **law of universal gravitation** but also the three laws of motion that we studied in Chapter 2. Some consider it the most important scientific book ever written.

law of universal gravitation Every object in the universe attracts every other object with a force that is directly proportional to the product of their masses and inversely proportional to the square of the distance separating them.

The gravitational law is beautiful in its simplicity. Everything pulls on everything else in a simple relationship that depends only on mass and distance.

*Every object in the universe attracts every other object with a force that is directly proportional to the product of their masses and inversely proportional to the square of the distance between their centers (***Figure 3.2***).*

$$\text{Force} = \text{constant} \times \frac{\text{mass}_1 \times \text{mass}_2}{\text{distance}^2}$$

$$\text{In symbols,} \quad F = G\frac{m_1 m_2}{d^2}$$

where G is the universal gravitational constant, equal to 6.67×10^{-11} N m^2/kg^2.

Gravitational force pair Figure 3.2

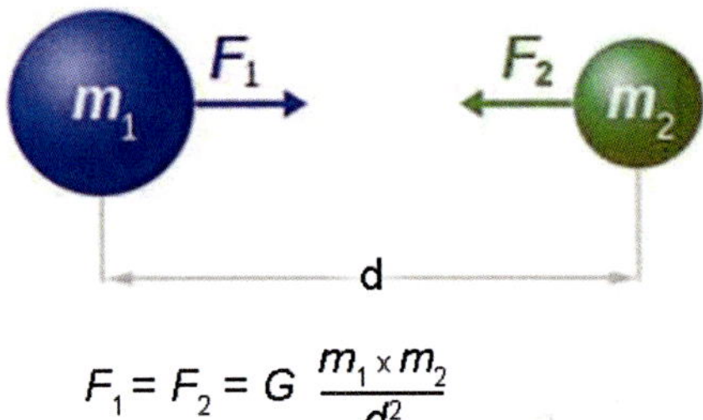

The gravitational force one object exerts on a second object is directed toward the first object and is equal and opposite to the force exerted by the second object on the first.

But Newton did not know the proportionality constant G and only tried to estimate it. Therefore, he could not use his equation to actually calculate the force between two objects. The small size of this constant made it very difficult to measure in a regular laboratory setting, where only small masses are available to use in experiments. More than a hundred years later, long after Newton's death, Henry Cavendish measured G in 1798. It was a very clever but difficult experiment. See **Figure 3.3**.

Cavendish experiment to determine G Figure 3.3

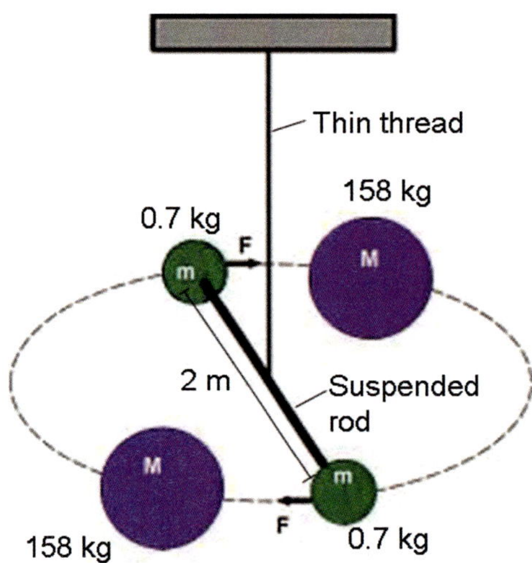

Henry Cavendish measured the extremely small force of gravitational attraction between two small lead spheres on a rod 2 m long and much larger lead spheres placed near them. Then he calculated G, the gravitational constant, which was very near today's accepted value of 6.67×10^{-11} N m^2/kg^2.

The gravitational force becomes significant only when at least one of the two masses is large. Two friends, even massive friends, do not feel their mutual gravitational attraction when standing next to each other, as **Example 3.2** shows.

Example 3.2
Gravitational force between two elephants
What is the gravitational force between two elephants? What is the gravitational force between one elephant and Earth?

Let's assume the elephants' centers are about 5 m apart and their masses are 5000 kg each. We apply Newton's Gravitational Law:

$$\text{Force} = G\frac{m_1 m_2}{d^2} = 6.67 \times 10^{-11}\,\text{N m}^2/\text{kg}^2\,\frac{(5000\text{ kg})(5000\text{ kg})}{(5\text{ m})^2} \simeq 6.7 \times 10^{-5}\,\text{N}$$

The attractive force between the elephants is truly small. This force makes the weight of a 0.1-gram head of a pin about fifteen times larger in comparison.

Let us see what this force is between one of the elephants and the Earth. We need to know the distance from the surface to the center of the Earth and the mass of the Earth. The distance to the center of the Earth is about 6370 km and its mass is 5.98×10^{24} kg.

$$\text{Force} = G\frac{m_1 m_2}{d^2} = 6.67 \times 10^{-11}\,\text{N m}^2/\text{kg}^2\,\frac{(5000\text{ kg})(5.98 \times 10^{24}\text{ kg})}{(6.37 \times 10^{6}\text{ m})^2} \simeq 49{,}000\,\text{N}$$

This is the same force as the weight of the elephant, which you already know from

Weight = mg = (5000 kg)(9.8 m/s^2) = 49,000 N.

When we calculate the gravitational force on any object on the surface of the Earth we use the above expression for weight because the gravitational acceleration g = 9.8 m/s^2 already contains all the information about G, mass of the Earth, and Earth's radius that we needed to calculate the force using the law of universal gravitation.

We have examined the influence of the size of the product of the two masses on the gravitational force between them; let us examine the distance dependence now. The strength of the gravitational attraction of any object to the Earth decreases as $1/d^2$, where d is the distance of the object to the center of the Earth. Thus, if you are one Earth radius above Earth's surface, the gravitational force will drop to $1/2^2$, or 1/4 of its size on the surface. See **Figure 3.4**. The acceleration due to gravity there would be about 2.5 m/s^2. There is a small variation on the surface of the Earth as well. The Earth's radius is about 21.5 km greater at the equator than at the poles. This variation is only about 0.1%.

Gravitational force versus distance Figure 3.4

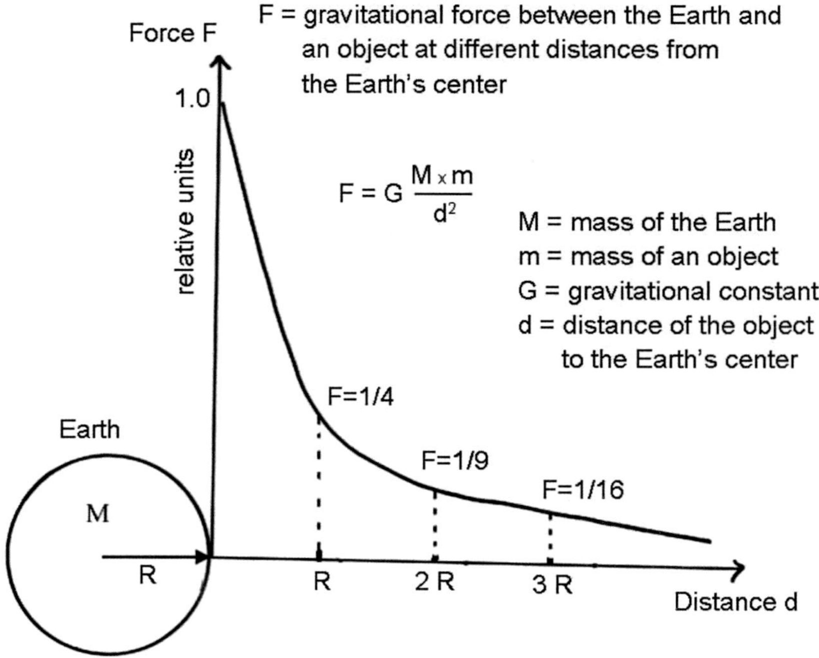

Gravitational force is a universal force. To the best of our knowledge, all massive objects in the Universe exert gravitational forces on each other. The law of universal gravitation explains the motion of all planets in the Solar system, including the irregularities due to mutual attraction of all the other planets. Astronomers have confirmed that gravitational force exists between stars in our own Milky Way Galaxy and in the neighboring Andromeda Galaxy.

CONCEPT CHECK

1. **How** would the gravitational force on any object on the surface of Earth change if Earth doubled its mass while keeping its size?

3.2 Laws of Planetary Motion

LEARNING OBJECTIVES

1. **State** Kepler's laws.

Newton's discovery of the law of universal gravitation was preceded by the work of a couple of astronomers, Tycho Brahe (1546-1601) and Johannes Kepler (1571-1630). Brahe, a distinguished Danish astronomer, spent over 20 years painstakingly recording positions of planets with huge, brass protractor-like instruments. His measurements proved very accurate even though he made them without a telescope, which had not been invented yet. Kepler started as Brahe's assistant and took over after Brahe died in 1601 in Prague. Brahe left behind a wealth of data that Kepler spent many years analyzing. This effort led him to a planetary model that was Sun-centered and supported Copernicus's and Galileo's efforts to reject the Earth-centered model of the ancient astronomer Ptolemy.

Kepler's Laws

First, Kepler studied the shape of the planetary orbits. He was looking for circular orbits with the Sun at the center. But he was disappointed. Brahe's data did not support that assumption and showed that they were elliptical. **Figure 3.5** describes an ellipse.

Ellipse Figure 3.5

You can draw an ellipse by fixing a string at two points (foci) and moving a pencil around the path permitted by the string.

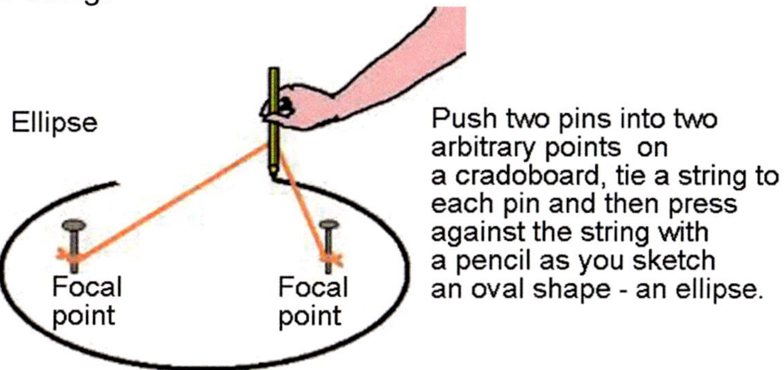

In the first of his three laws Kepler stated: *The planets move along elliptical orbits rather than circular ones, with the Sun at one focus.*

Kepler also noticed that the planets do not revolve around the Sun at a constant speed but move faster when they are nearer the Sun and slower when they are farther from the Sun. This was something he could not explain; the explanation had to wait until Newton showed that the centripetal force holding a satellite or a planet in an orbit decreases with the inverse of distance squared, and therefore the orbital speed must decrease too. This explained why satellites or planets further away from their orbital centers move at slower speeds as is shown by a simple derivation below

$$\frac{mv^2}{r} = G\frac{Mm}{r^2} \text{ or } v^2 = \frac{GM}{r}.$$

Kepler's observation of varying speeds led him to compare the areas swept out by a line joining the Sun and the planet. What resulted was Kepler's second law: *Each planet moves in such a way that a line joining it and the Sun sweeps out equal areas of space in equal time intervals.* See **Figure 3.6**.

Kepler's second law Figure 3.6

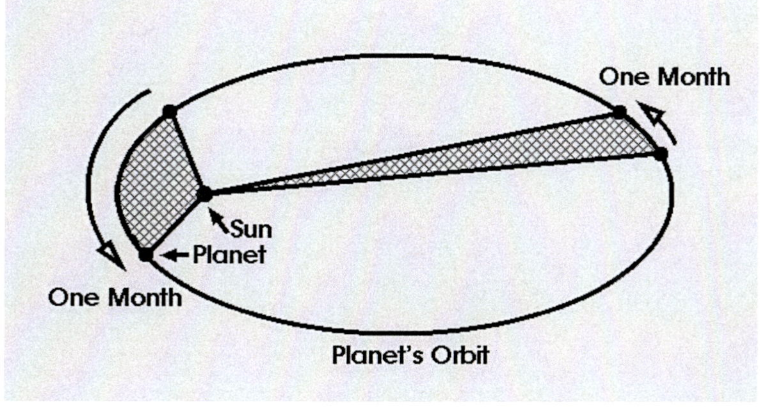

There was something else that Kepler wanted to find and that proved elusive. He searched for the connection between the time it takes a planet to orbit the Sun and its distance from the Sun. After ten years of intensive search through the data, Kepler discovered the third law: *The square of the orbital period T is proportional to the cube of the average orbital radius r (or distance) of the planet from the Sun.*

If Kepler knew Newton's laws, he could had saved himself lot of time by this simple derivation:

$$T = \frac{2\pi r}{v} = \frac{2\pi r}{\sqrt{\frac{GM}{r}}} \text{ or } T^2 = \frac{4\pi^2}{GM} r^3 .$$

This relation means that the ratio of T^2/r^3 is the same for all planets in the solar system and if one knows the period T, one can find the distance r or vice versa. The third law is also useful in calculating proper orbits or periods of satellites. For a satellite close to Earth, the period is about 90 minutes. At higher altitudes, the orbital speed is smaller and the period is longer. Communications satellites are in an orbit of about 5.5 Earth radii above the Earth's surface and orbit at the same rate as the rotation of the Earth, once every 24 hours.

Kepler's discoveries of the nature of planetary motion are known today as **Kepler's Laws**. They apply not only to planets in the solar system, but also to satellites or spacecraft such as the space shuttle orbiting the Earth, but here we replace the Sun with the Earth at one focal point of the orbit. Kepler's laws added to the accuracy with which one could predict the positions of planets as they appear in the sky around the fixed stars. The laws set the stage for Newton to explain the force that made the planets orbit the Sun and to formulate his universal gravitational law.

Kepler's laws 1. The planets move along elliptical orbits rather than circular ones, with the Sun at one focus. 2. Each planet moves in such a way that a line joining it and the Sun sweeps out equal areas of space in equal time intervals. Planets move in such a way that a line joining the Sun and the planet sweeps out equal areas of space in equal times. 3. The square of the orbital period T is proportional to the cube of the average orbital radius r of the planet from the Sun.

CONCEPT CHECK

1. **Why** does Earth speed up and slow down as it orbits the Sun?

3.3 Ocean Tides

LEARNING OBJECTIVES

1. Explain tides.

The level of water in our oceans continuously changes. The water line can move as much as 100 meters along a coastline in several hours. For centuries, port cities have kept records of the times and levels of high and low **tides**. They are caused by differences in the gravitational pull of the Moon on opposite sides of the Earth. See **Figure 3.7**.

tides Tidal bulges in the oceans caused primarily by the difference in the gravitational pull of the Moon on opposite sides of the Earth.

Elongation of the Earth's shape as a result of the Moon's gravitational pull Figure 3.7

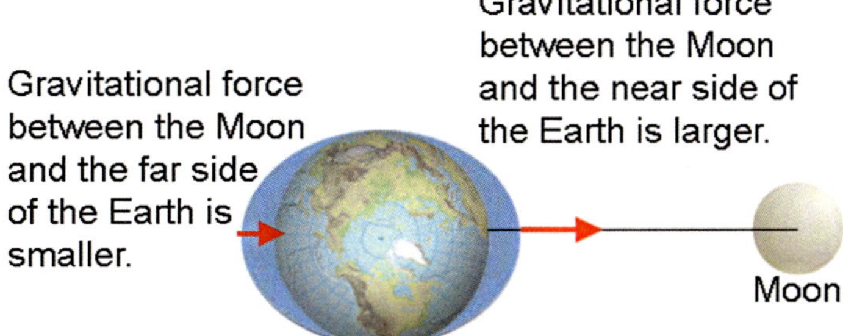

Water on the near side is racing toward the Moon,while the water on the far side is left behind as the Earth moves through space.

Think of the oceans (covering 70% of the Earth's surface) as being all connected and acting like a ball made of Jell-O. If you pull harder on one side than the other, the ball becomes elongated. If the surface of the water goes down somewhere, it has to rise somewhere else. The effect of all these forces is to stretch the oceans out a tiny bit along the straight line between the centers of the Earth and Moon. The water piles up at opposite sides of the Earth. These piles of water are called the tidal bulges, and as the Earth rotates, points on its surface move into and then out of the bulges, resulting in two high tides that "travel" around the Earth daily. One quarter turn after each high tide the Earth rotates through a low tide, or local water depression (**Figure 3.8**).

High and low tides Figure 3.8

Earth rotates in and out of high and low tides.

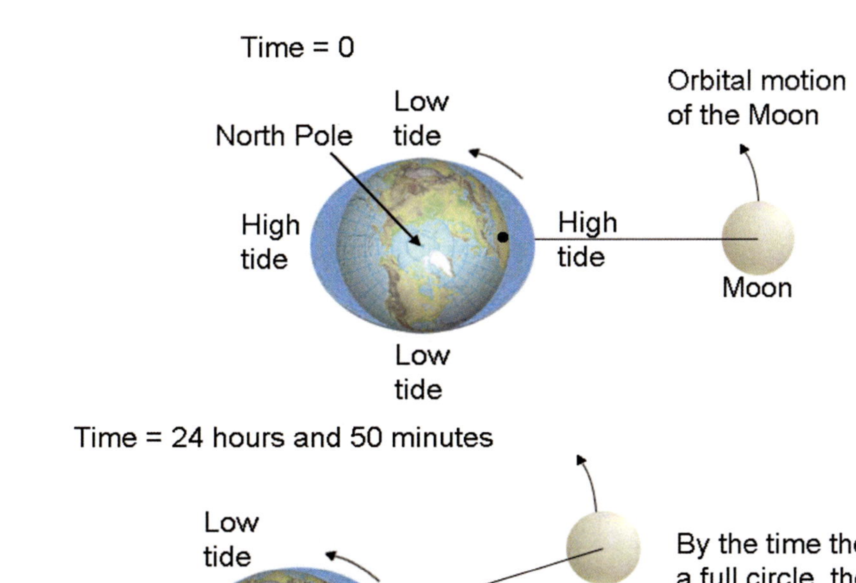

As the Earth turns, the Moon continues farther along in its orbit around the Earth, so a given location on the Earth needs to turn a little extra to catch up with the orbiting Moon again. The Earth's rotation relative to the Moon is 24 hours and 50 minutes. The high tides are 12 hours and 25 minutes apart, as are the low tides.

In addition to the Moon, the Sun also exerts a gravitational pull on the Earth. So we might expect to see solar tides too. However, the Sun's contribution to tides is smaller than the Moon's, even though the Sun's net gravitational pull on the Earth is far stronger. This is because the percentage difference in the Sun's gravitational force from one side of the Earth to the other is less than the Moon's, and it is the larger percentage difference in the gravitational force from one side of the Earth to the other that causes larger tides. Similarly, you will not see tides on a lake because the lake is too small to feel different gravitational pull on each of its opposite ends.

The calculation of a tide table is a challenging problem because the Moon, Earth, and Sun, are continuously moving. The position of the Moon relative to the Earth and the Sun plays the most important role.

CONCEPT CHECK

1. **What** is the position of the Moon relative to Earth and the Sun when the tide is highest?

Summary

3.1 Gravitational Force

- **Law of Universal Gravitation**
  Every object in the universe attracts every other object with a force that is directly proportional to the product of their masses and inversely proportional to the square of the distance between their centers.

$$\text{Force} = \text{constant} \times \frac{\text{mass}_1 \times \text{mass}_2}{\text{distance}^2}$$

In symbols, $$F = G\frac{m_1 m_2}{d^2}$$

where G is the universal gravitational constant equal to 6.67×10^{-11} N m^2/kg^2.

3.2 Laws of Planetary Motion

- **Kepler's laws**
  First law: Planets follow elliptical orbits.
  Second law: A line joining the Sun and a planet sweeps out equal areas of space in equal time intervals.
  Third law: The square of the period is proportional to the cube of the radius. All planets in the solar system obey this law.
- **Ocean Tides**
  The tidal bulges in the oceans are caused primarily by the difference in the gravitational pull of the Moon on opposite sides of the Earth. As the earth rotates, points on its surface move from low **tide** to high tide, back to low tide, and so on.

Key terms
weightlessness
law of universal gravitation
Kepler's laws
tide
spring tides
neap tides

Critical Thinking Questions

1. How far away from the surface of the Earth do you have to be to experience a gravitational attraction one-ninth the size of what it is on the Earth?
2. We seem to have two different formulas for the force due to gravity. One calculates your weight and the other calculates your gravitational attraction to the Earth. Explain.
3. Would you expect the value of gravitational attraction to be larger or smaller at the equator, compared to its value at the Earth's poles?
4. If a satellite in an orbit around the Earth is continuously falling, why is it that it does not crash into the Earth?
5. If an object is in a circular orbit around the Earth, its speed remains constant and is about 8 km/s. Why doesn't the gravitational force acting on it change its speed?
6. Is it possible for an Earth satellite to be stationary over a city? If so, what law would you apply to find its proper orbital distance?
7. Is the high tide at a given location always at the same time? If not, how may it vary from day to day?

Exercises

1. A satellite is placed in a circular orbit to observe the surface of Mars from an altitude of 144 km. The equatorial radius of Mars is 3397 km. If the tangential speed of the satellite is 3480 m/s, what is the magnitude of the centripetal acceleration of the satellite?
2. Show that if you calculate your weight using Newton's gravitational law $F = GmM/d^2$, you should get about the same answer as that obtained using $w = mg$. Assume your mass is 70 kg. The mass of the Earth is 6×10^{24} kg and the distance to the center of the Earth is about 6370 km.
3. Two objects in outer space attract each other with a gravitational force of 1000 N; if their distance apart is halved, how does the force change?
4. What is the attractive force between two 220-lb opposing football players positioned about 1 meter apart? Note that a weight of 220 lb is equivalent to a mass of 100 kg. How does this compare to the weight of a baby flea that weighs about 10^{-6} N?
5. What is the gravitational force between the Earth and the Moon? The mass of the Earth is about 6×10^{24} kg, the mass of the Moon is about 7.3×10^{22} kg, and their mean distance apart is about 3.9×10^{8} m.

Chapter 4
Energy and Momentum

Energy is the central and perhaps most important concept across all scientific fields, not just in physics. Understanding energy and the development of energy resources is critical to our survival. Energy comes from numerous sources and takes many different forms. All of the radiation and all of the motion we see in the Universe are forms of energy. Energy is constantly changing from one form to another and is behind everything that happens.

In addition to studying energy in this chapter, we will examine momentum, which is a property of all moving objects. Both energy and momentum obey conservation laws, which allow physicists to determine the outcomes of many physical events, from the collisions of subatomic particles to the interactions of stars and galaxies. A boxer here illustrates both energy and momentum concepts. We will show how he translates force over a distance into energy at impact and how the stopping time of the glove in contact with the opposing boxer's face influences the size of the force acting on him.

4.1 Work, Energy, and Power

LEARNING OBJECTIVES

1. **Define** mechanical work.
2. **Describe** forms of mechanical energy.
3. **Distinguish** the difference between power and work.

Without energy, there would be nothing—no Sun, no wind, no rivers, no oceans, no life at all. All living things need energy for growing, moving, keeping warm, and other activities. Our civilization and improving living standards have developed parallel to our ability to generate and use energy.

Even early humans and civilizations relied on a combination of energy sources. Stone Age cave dwellers used fire to cook food, light their caves, and keep warm. They also developed tools to convert energy from their muscles more effectively so that they could lift objects, grind grain, push plows, or transport people and goods. Later, energy from fire was used to smelt ore, fire pottery, and make glass. Viking ships used human muscle energy by use of oarsmen, and wind energy when conditions were good.

Work

Work and energy are closely related, and understanding work will help us define energy. In our daily lives we say we have to do work to accomplish some task. After a day's work, we feel tired because we expended energy doing work. Yet the meaning of work for a physicist may be different from what you might consider work. For instance, standing in an airport, holding a heavy suitcase in your hand, you may feel that you are doing work. A physicist will tell you that you are performing no mechanical work on the suitcase because work involves both force *and* motion. How does a physicist define work?

work The product of the force exerted on an object and the distance moved by the object

When you push on a crate, causing it to slide some distance, you perform work. The heavier the crate and/or the farther you push it, the more work you do. While doing work, you are transferring your energy to the crate. In general, when work is done by one object on another, energy is transferred between the two objects and work is what transfers it. The more force you apply and the farther you move the object, such as a crate or a stalled car, the more mechanical work you do. See **Figure 4.1**.

Work done on a stalled car Figure 4.1
People pushing on a stalled car transfer their energy to the car and do work on it. The work they do is the product of their force and the distance the car moves.

Distance moved d

Work is the product of the force $F_{//}$ in the direction of motion and the distance moved d.

$$\text{Work} = \text{Force} \times \text{distance}$$

In symbols,

$$W = F_{//} \times d$$

If the force is measured in newtons and distance in meters, the units of work are joules (J). One joule is the amount of work done by a force of 1 N acting through a distance of 1 m. When forces are acting on an object in directions other than along the displacement of the object, then only the component of the force along the direction of motion contributes to work.

When work is being done, other forces may be present that oppose the motion. In other words, you may do work on something even though you force it to move against an opposing force. For instance, when you lift a stack of books, work is required to raise the books against the force of gravity. Or when you push on a stalled car, you move the car against an opposing force of friction. In these situations we say that we are doing work against something—against the force of gravity or friction, respectively. We should always specify the force that does work because just saying that work was done may not be clear in the presence of several forces.

Mechanical energy

With the concept of work, we can describe **energy**. The energy of a system is the amount of work it can do, regardless of whether or not it actually does work. By a system, we mean a particular collection of objects or a specific part of the universe. Work and energy, even though they share the same units, are not the same thing. Work is a process, while energy is a property of a system.

energy The property of a system that allows it to do work

Energy can take on many forms, some of which we will examine later in this chapter. Here, we will focus on the most common form—mechanical energy. Mechanical energy is energy due to the position of something or movement of something; it is stored as **potential energy** or **kinetic energy**, respectively.

potential energy Energy of a system due to its position

kinetic energy Energy due to motion

There are many different forms of potential energy. These include *gravitational potential energy*; *chemical potential energy* (contained in fuels such as gasoline or the foods you eat); *nuclear potential energy* (present inside the nucleus); *elastic potential energy* (for example, the energy of a compressed spring); and *electromagnetic potential energy* (such as the energy of the Sun's radiation). We will focus here on gravitational potential energy and elastic potential energy.

A diver standing still at the edge of a diving board has energy—*gravitational potential energy*. She gained this energy by climbing up the steps to the top of the diving platform. Her gravitational potential energy is determined by her mass m, the gravitational acceleration due to gravity g, and her vertical position—her height h. Height can be measured from the ground, the floor, or some other reference level. We often define this reference level so that the gravitational potential energy is zero there. In the case of the diver, we define the water's level in the pool as the zero reference level.
The gravitational potential energy of an object is calculated by multiplying its mass m by the acceleration due to gravity g and its height h.

Gravitational potential energy = mass × acceleration due to gravity × height

In symbols,

$$PE = m \times g \times h$$

Doubling either the mass of the object or its height doubles its gravitational potential energy. Therefore, if the diver climbs to a higher height, she will increase her potential energy as well (**Figure 4.2**). **Example 4.1** shows how to find the potential energy of a pole-vaulter.

Gravitational potential energy of a diver Figure 4.2

A diver's mass *m*, her height *h* above the water level, and the gravitational acceleration due to gravity *g* determine her gravitational potential energy.

Example 4.1
Gravitational potential energy of a pole-vaulter
A world-class pole-vaulter can clear a bar 6.00 m high. What is his potential energy at the top of the bar if his mass is 78 kg?

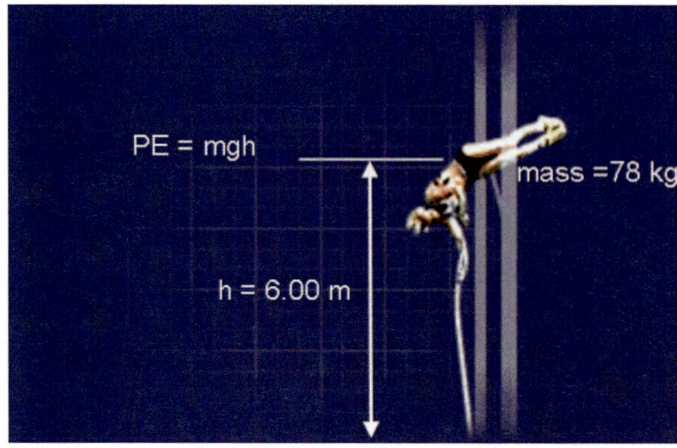

Solution:
We calculate the gravitational potential energy of the pole-vaulter by multiplying his mass *m* by the acceleration due to gravity *g* and his height *h* above the ground.

$PE = m \times g \times h$ = (78 kg)(9.8 m/s^2)(6.00 m) = 4586.4 J.

He gains 4586.4 joules of potential energy at the top of the bar.

Another form of potential energy is that of an object, such as a spring, that is stretched or compressed. We call this potential energy *elastic potential energy*. An object is elastic if it returns to its original shape after it is compressed or stretched. By stretching a spring or a rubber band you do work on it and it is stored as potential energy. This stored energy can be used to do work. If you place an object in front of the compressed spring and let go of it, the spring will bounce back and its potential energy will convert to kinetic energy of the object. A stronger spring or a larger compression of the spring would store more potential energy, so it would transfer more energy to the object.

Many more examples of elastic potential energy occur in our daily lives. For instance, the shock absorbers on your bike or your car can store a sudden jolt of energy that your vehicle receives when it drives over a hole in the road. The rider does not feel the full impact of this energy, so the ride feels more comfortable.

When an archer releases an arrow it gains kinetic energy (**Figure 4.3**). Kinetic energy (*KE*) represents the energy of motion. The kinetic energy of a moving object such as the arrow depends only on its mass and speed. We define it as

$$\text{Kinetic energy} = \frac{1}{2}\text{mass} \times \text{speed}^2 .$$

Elastic potential energy of an arrow Figure 4.3

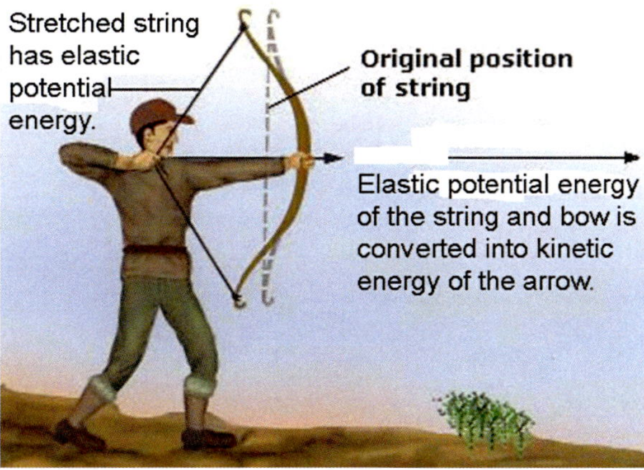

In symbols,

$$KE = \frac{1}{2}mv^2.$$

Notice that doubling the mass of an object would double its kinetic energy, but doubling the speed of an object would quadruple its kinetic energy, since kinetic energy is proportional to the square of an object's speed.

To find the kinetic energy in joules, we insert into the formula the values for mass in kilograms and speed in meters per second.

Power

We have learned about energy and work, but we have said nothing about how long it takes us to do this work or to convert energy from one form to another. When we lift a stack of books higher, we do the same amount of work whether we do it quickly or slowly. But if we do it quickly, we become more tired. The difference is *the rate of doing work.* Doing work at a faster rate requires more **power**. To increase power, you can do a given amount of work in less time or you can increase the amount of work done in a given time.

power Time rate of doing work

Power is the rate at which energy is used; it is equal to the amount of work done per time interval, or

$$\text{Power} = \frac{\text{Work}}{\text{Time}}$$

In symbols,

$$P = \frac{W}{t}$$

Power can be also thought of as energy produced or consumed divided by the time taken. The units of power are joules per second, also defined as watts (W). Another common unit is the kilowatt or kW, where one kilowatt equals 1000 watts. In the United States we often rate engines in *horsepower* units, where one horsepower is about three-fourths of a kilowatt. Rearranging the equation for power, we see that energy is power times time ($P \times t$). Therefore, the kilowatt-hour (kWh) is a unit of energy. This is the unit that electric power companies use when they charge for electricity.

Many interesting examples of power consumption arise in our lives. Just sitting quietly, you radiate thermal energy at a rate of 75 W. In that sense, you are like a 75-W light bulb. A 70-kg (154 lb) man normally uses about 10^7 J/day. This computes into an average energy consumption rate (known among doctors as the metabolic rate) of about 120 W.

A long jumper can generate as much as 8000 watts at take-off. The same athlete could maintain a power output of about 400 watts if performing for an hour. **Example 4.2** calculates the power output of an Olympic weightlifter.

Example 4.2
Power output of an Olympic weightlifter
An Olympic weightlifter lifts 210 kg above his head, 2 meters above the floor, in 0.9 second. What power does he need to accomplish this feat?

Solution:
Power is the rate at which energy is used. It is equal to the amount of work done per time interval, or

$$\text{Power} = \frac{\text{Work}}{\text{Time}}$$

The weightlifter's work is equivalent to the increase in potential energy of the weight.

Potential energy (*PE*) = mass m × acceleration due to gravity g × height h.

Therefore,

$$PE = mgh = (210 \text{ kg})(9.8 \text{ m/s}^2)(2 \text{ m}) = 4116 \text{ J}.$$

His power output is then

$$\text{Power} = \frac{4116 \text{ J}}{0.9 \text{ s}} = 4573 \text{ W}$$

This is an impressive number. However, he cannot sustain this output for longer than a brief moment.

CONCEPT CHECKS
1. **What** component of force is used to define work?
2. **How** can you increase your kinetic energy and potential energy all at once?
3. **How** could you increase your power output while lifting a pile of bricks?

4.2 Energy Transformations

LEARNING OBJECTIVES

1. **Define** the work-energy theorem.
2. **Explain** the law of energy conservation.
3. **Describe** the different energy sources and energy forms.

It takes work to change the speed of an object and therefore its kinetic energy. For instance, when you push on an object and it speeds up, you are doing mechanical work; when you pull on a moving object and it slows down, you are doing mechanical work as well. In the process, energy is transferred to either increase or reduce the kinetic energy of an object. Work enables this transfer of energy. Work and the kinetic energy of an object are closely linked. We can determine this relationship quantitatively.

The Work-Energy Theorem

When you throw a ball, you do work on it and the ball leaves your hand with some speed. It has kinetic energy, which is equal to the work needed to bring it from rest to its final speed. In words,

Work = Net force on the ball × distance over which the force acts
= Final kinetic energy of the ball

In symbols,

$$F \times d = \frac{1}{2} m \times v^2$$

We can generalize from this conclusion and say that *work done on an object equals the change of its kinetic energy.*

In symbols,

$$\text{Work} = \Delta KE = KE_{\text{final}} - KE_{\text{intial}}$$

We call this relation the **work-energy theorem**. See **Figure 4.4** for an example.

work-energy theorem The work done on an object equals the change in its kinetic energy.

Work-energy theorem Figure 4.4

During a shot put, the hand applies force over some distance. The hand does work on the shot, causing an increase in the ball's kinetic energy.

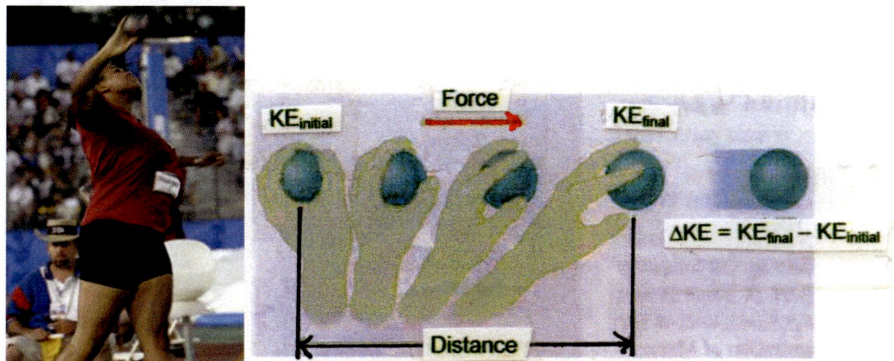

We need to emphasize that the work in the work-energy theorem is the *net* work—that is, work based on the net force. For example, if you push on a sliding crate and friction acts on it as well, then you need to find the net force acting on it (that is, your push minus friction) because only part of your force is acting to change the crate's kinetic energy. The rest of your force is dissipated by friction. See **Example 4.3**.

Example 4.3
Work done on a crate
A man pulls a stationary 35-kg crate across a rough horizontal floor by applying a constant horizontal force of 125 N for a distance of 4 m. A frictional force of 25 N opposes the motion. What is the crate's final energy?
Solution:
We can say that the net work done on the crate equals the change in kinetic energy of the crate.

$$\text{Net force} \times \text{distance} = \text{change in kinetic energy}$$

In symbols,

$$F_{net}\, d = KE = (125\ \text{N} - 25\ \text{N})(4\ \text{m}) = 400\ \text{J}.$$

The initial kinetic energy of the crate was zero because it was not moving, and we have just calculated the change in kinetic energy as 400 J. So the crate's final energy is 400 J.

Conservation of Energy

We have seen several examples of energy changing from one form to another. The total energy during these conversions remains unchanged. This is the result of one of the most important laws of nature—the law of **conservation of energy**.

conservation of energy Energy may be transformed into many different forms, but the total energy never changes.

Energy cannot be created or destroyed; it may be transformed into another form, but the total energy never changes.

In other words, the total energy of any isolated system is constant. The conservation of energy law may not always be apparent. For instance, when you stop pedaling while riding a bike on flat terrain, you come to a stop. What happened to your kinetic energy? You came to a stop because of frictional forces on the bike and also on specific moving components of the bike, such as ball bearings in the wheel that rotate with the wheel. Friction caused heating of the ball bearings or bike's tires, for instance. The bike's kinetic energy converted into thermal energy. This thermal energy subsequently dissipated into the environment. If you consider the bike and the environment as one isolated system, the total energy in this system remains constant.

A falling object provides another example of conservation of energy. As the object falls, its gravitational potential energy decreases. The potential energy does not just vanish, but is converted to kinetic energy. During the fall, the object's kinetic energy increases while its potential energy decreases. The sum of the potential and kinetic energies is the total *mechanical energy E* and is constant.

Total energy = kinetic energy + gravitational potential energy = constant

In symbols,

$$E = KE + PE = \text{constant}$$

Here we neglect any frictional losses due to motion in air. See **Figure 4.5**. Mechanical energy is not conserved if a dissipative force is present; however, the total energy, including the dissipative energy losses, is conserved. **Figure 4.6** describes a snowboarder's transformation of kinetic energy to potential energy, and **Example 4.4** examines numerically the conservation of energy during a bobsled run.

Conservation of mechanical energy Figure 4.5

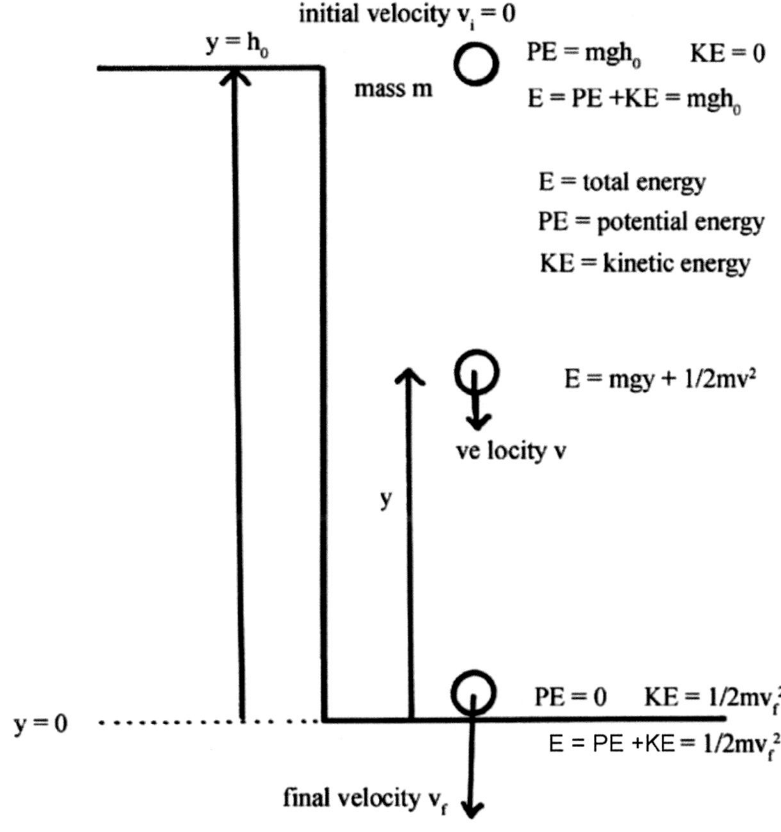

Total energy is conserved for an object falling to the ground. Total energy of the rock at the top (PE) is equal to its total energy at the bottom (KE) just before it hits the ground.

$$mgh = \frac{1}{2}mv_f^2$$

Solving for v_f, we get

$$v_f = \sqrt{2gh_0}$$

Applying the conservation of energy, we discovered the final velocity of the rock just before it hits the ground.

Conversion of kinetic energy to potential energy and back Figure 4.6

a. A half-pipe skier or snowboarder converts most of her initial kinetic energy KE (1) to potential energy PE at the apex of her motion (2). This potential energy is then converted back to her initial kinetic energy (3) if there are no frictional losses.

b. The motion of the skier/snowboarder is similar to the motion of a pendulum whose energy oscillates back and forth between the kinetic and potential energies. If there are no frictional losses, then the sum of kinetic and potential energies is constant and the oscillation continues undiminished.

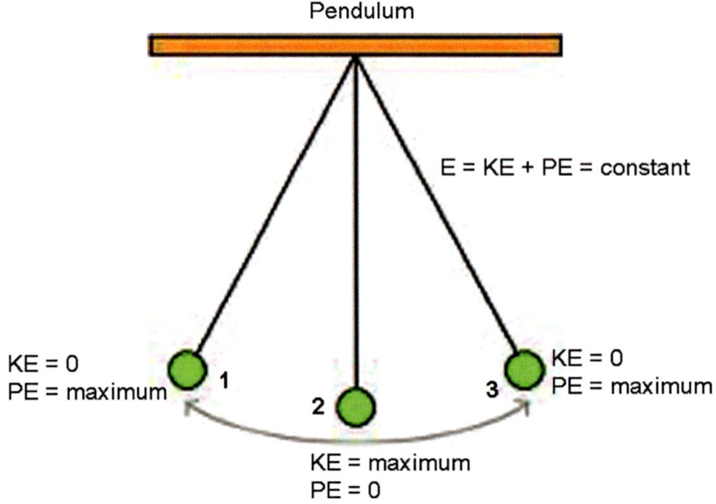

c. Kelly Clark, gold medal winner in the women's halfpipe at the 2002 Salt Lake City Olympics, was able to consistently catch "big air" on her tricks compared to her competitors, who lost "air" on their subsequent tricks.

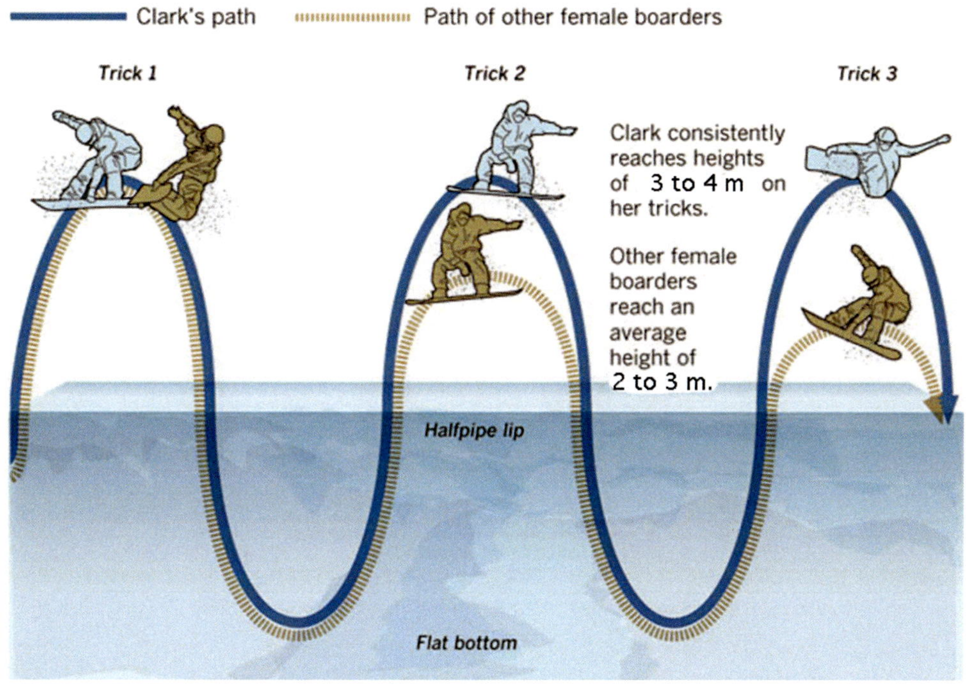

Critical thinking…
Why do you think the other boarders reached lower heights on each successive trick?

Example 4.4
Conservation of energy during a bobsled run
A bobsled team starts a run on a Lake Placid track that has a height difference between the start and finish of 107 m. The mass of the bobsled with a crew of four is 630 kg. Assuming no energy loss due to friction, what is the loaded bobsled's kinetic energy and speed at the finish line? (a) First, assume a standing start. (b) Then solve again assuming the initial kinetic energy of the crew and bobsled is 15,000 J.

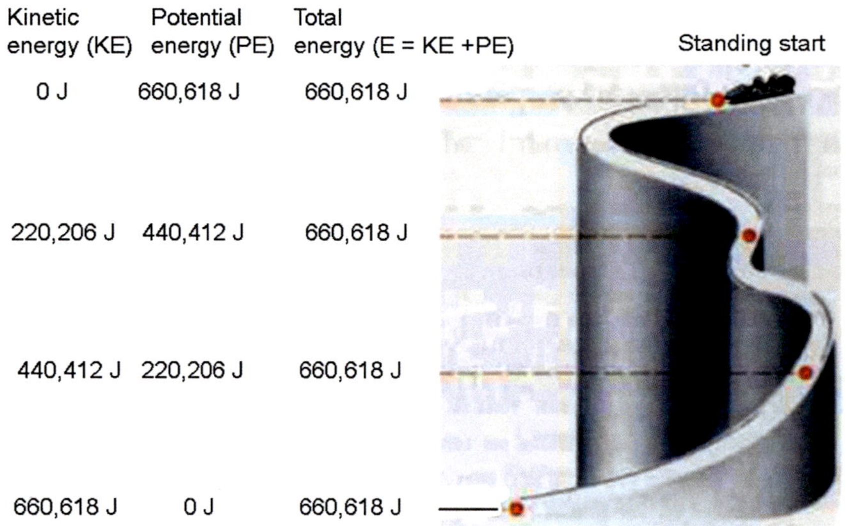

Solution:
(a) At the top of the run, the bobsled's energy is the sum of its kinetic energy (KE) and gravitational potential energy (PE). We know that

Gravitational potential energy = mass $\times$ acceleration due to gravity $\times$ height or

$$PE = m \times g \times h = (630 \text{ kg})(9.8 \text{ m/s}^2)(107 \text{ m}) = 660{,}618 \text{ J}$$

This is the bobsled's total energy, which is gradually transformed into kinetic energy, as the bobsled travels down the track. At the bottom, the bobsled's energy is all kinetic and therefore it must be 660,618 J as well.

(b) If the bobsled team has a running start, we need to add the initial kinetic energy to the potential energy at the top to find the bobsled's total energy. The total energy at the top is therefore

$$E = KE + PE = 15{,}000 + 660{,}618 = 675{,}618 \text{ J}.$$

This is the bobsled's kinetic energy at the finish line. The running start allows it to reach higher velocity and get to the finish faster.

We are constantly consuming or producing energy in our lives. In the United States, the total rate of usable energy consumption is about 10,000 W per person. This is the same as 10,000 J per second, or 846,000,000 J of energy per day. Most of this energy is used to increase our comfort level.

Energy can take on many different forms. Let's follow the energy transformation during the operation of a car, for instance. All the energy that a car uses to accelerate and keep moving (kinetic energy) comes from the gasoline or diesel fuel (chemical potential energy) it burns in its engine. If you suddenly step on the brakes to stop, the kinetic energy of the car decreases to zero. As the brake pads slow down the car wheels, they convert the energy of motion into heat. The brake pads and the wheels heat up. This heat is released into the environment as radiation energy.

Energy comes from many different sources. The origin of fossil fuels—our most important energy source today—dates as far back as 300 million years ago, when plants in ancient forests died and were buried. Under the right conditions, they were fossilized and over millions of years turned into coal, oil, or natural gas. We extract energy from fossil fuels by burning them. Once burned, fossil fuels cannot be used again: they are nonrenewable. If we do not develop other energy sources soon, humankind will have problems once these sources run out.

CONCEPT CHECK

1. **How** is work done on a car, either by its engine or by people pushing on it, related to its kinetic energy?
2. **Why** does it appear that mechanical energy is not conserved when a moving object comes to a stop?
3. **What** renewable energy sources appear most promising for the future?

4.3 Momentum and Impulse

LEARNING OBJECTIVES

1. **Define** momentum.
2. **Describe** impulse and its link to momentum.

In Chapter 2 we introduced the concept of inertia, which is the property of an object that causes it to resist motion, and linked it to mass. An object has inertia whether it is at rest or in motion. Here, we study the inertia of moving objects. We will describe an object's motion by combining its mass and velocity to define a new term, **momentum**.

momentum The product of the mass of an object and its velocity

Defining Momentum and Impulse

Combining information about the mass and velocity of an object gives us a better way of comparing the motion of objects. For instance, a large freight truck and a car travelling at the same speed have different amounts of momentum simply because the truck has a larger mass. It would be more difficult to change the motion of the truck than to change the motion of the car. However, two cars with the same mass also have different amounts of momentum if they are moving at different speeds. It is more difficult to change the direction of the faster-moving car.

Momentum (plural *momenta*) is the product of mass and velocity,

$$\text{Momentum} = \text{mass} \times \text{velocity}$$

$$p = mv.$$

Momentum is a vector and points in the direction of velocity. Objects with a small mass and large velocity can have just as much momentum as objects having a large mass and small velocity, as shown in **Figure 4.7.**

Comparing momenta Figure 4.7

A fired bullet may have more momentum than a moving baseball because the bullet's velocity is significantly larger than that of the baseball, even though the bullet has a smaller mass.

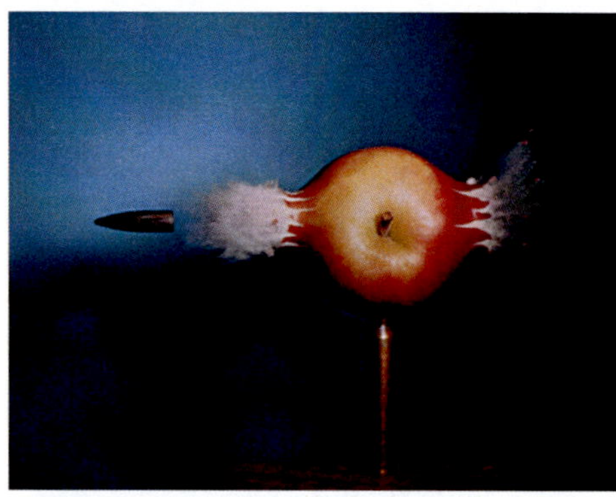

Bullet of mass $m = 0.068$ kg traveling with velocity $v = 175$ m/s has momentum p:

$p = mv = (0.068 \text{ kg})(175 \text{ m/s})$
$= 11.9$ kg m/s.

Baseball of mass $m = 0.145$ kg traveling with velocity $v = 31$ m/s has momentum p:

$p = mv = (0.145 \text{ kg})(31 \text{ m/s})$
$= 4.5$ kg m/s.

To change the momentum of any object, we need force to change the object's velocity. Something else is also important in changing momentum—how long the force acts. If you apply the force for a short period of time, a small change in momentum results. If you apply the force for a longer time, a greater change in momentum occurs.

In addition to changing how long the force acts, you can also vary the magnitude of the force. For instance, to stop a car, you can brake quickly or you can brake gently over a longer time and get the same change in momentum. In another words, a large force applied over a short time gives you the same result as a small force applied over a longer time. We can describe this quantitatively by recalling Newton's second law:

$$\text{Net force} = \text{mass} \times \frac{\text{change in velocity}}{\text{time}}$$

If we multiply both sides by time we get

$$\text{Net force} \times \text{time} = \text{mass} \times \text{change in velocity} = \text{change in momentum}$$

The quantity on the left, *net force* $\times$ *time,* is called **impulse**.

impulse The product of net force acting on an object and the time during which it acts

In symbols,

$$Ft = \Delta p = p_{\mathrm{f}} - p_{\mathrm{i}}$$

where Δ(delta) means "change in" and p_{f} and p_{i} are the final and initial momenta of the object. F represents the average net force and t represents the time interval during which the force acts.

The Impulse-Momentum Relation

The equation we used to define impulse is also known as the **impulse-momentum relation**. In some respects, it is similar to the work-energy theorem, in that it has a wide range of applications to many common situations.

impulse-momentum relation Impulse acting on an object is equal to the resulting change in momentum of the object.

This relation tells us that a car that is twice as massive as another needs twice as much impulse to stop it. One way of doing it is by applying twice as much force over the same time; another way is to keep the same force but apply it over twice the time. **Figure 4.8** describes the impulse-momentum relation of a tennis ball hit by a tennis racket.

Impulse of a tennis ball hit by a racket Figure 4.8

a. If you hit a tennis ball with a tennis racket, the net force on the ball increases at first, as the ball's shape becomes deformed. This force on the ball then decreases as the ball gradually regains its original shape and leaves the racket.

b. Even though the net force on the ball is changing during its collision with a tennis racket, in our description we use the average net force to keep the math simple.

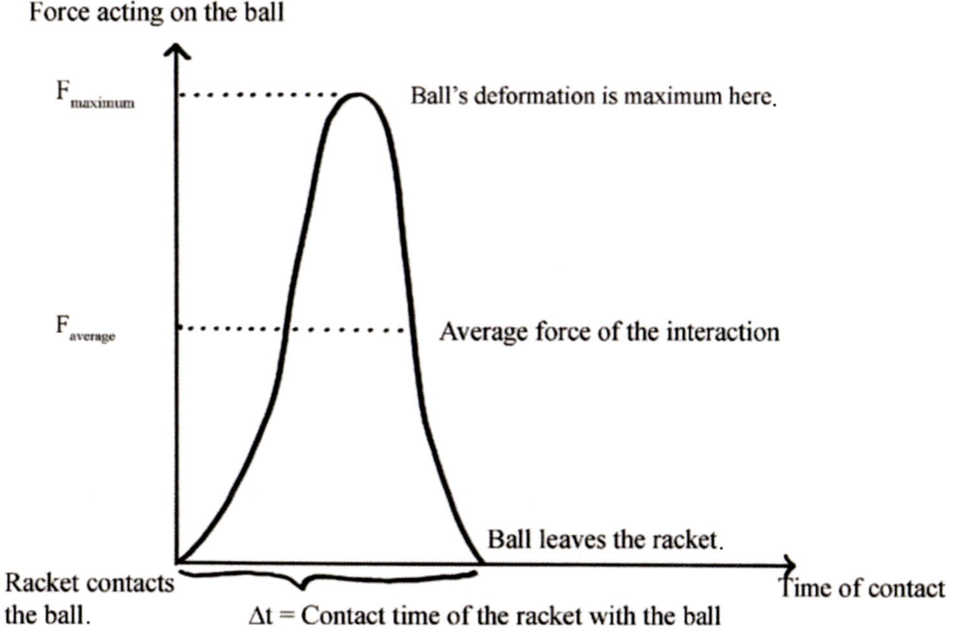

We can apply the impulse-momentum relationship during landing when jumping from significant heights to learn how to reduce the force on our feet. See **Figure 4.9**. The change of momentum is the same no matter what surface one lands on. However, a softer landing surface has a longer stopping time because grass has a bigger give, a little like a soft mattress. Therefore, the average net force of the ground acting on the feet is smaller than when you land on concrete.

Impact force acting on feet during landing Figure 4.9

Softer colliding surfaces help prevent injuries in other applications, such as airbags in cars or soft boxing gloves. Both the airbags and the soft boxing gloves increase the collision time and therefore reduce the average force on your face. **Example 4.5** shows how you can apply the impulse-momentum relation to calculate the force that a karate expert generates to break a stack of boards.

Example 4.5
Force needed to break a stack of boards
What average force does a karate black belt master apply with her hand to break a stack of 8 wooden boards? The effective mass of her hand and forearm is 3 kg. Videotape showed her hand swinging through 1/2 meter in the time between two frames (1/30 second). The time of this interaction was 0.02 seconds.

Solution:
When she brings her hand to rest, the change in momentum of her hand, and therefore its impulse, are the same no matter how long it takes for the hand to come to a stop. But if the collision contact time is short, the impact force will be large. That is why the striker moves her hand very quickly and makes the contact time with the boards very brief. This allows her to impart a large force.

Since the videotape showed the hand swinging through 1/2 meter in the time between frames (1/30 sec), the velocity of her hand before the collision was 15 m/sec. Now we can apply the impulse-momentum relationship.

(Net force)(time) = change in momentum, or

(Net force)(0.02 s) = (3 kg)(15 m/s) = 45 kg m/s.

$$\text{Net force} = \frac{(3\text{ kg})(15\text{ m/s})}{0.02\text{ s}} = 2{,}250\text{ N}$$

Therefore, the net force is 2,250 N! This is equivalent to over 504 pounds!

CONCEPT CHECKS

1. **How** could a flying mosquito have greater momentum than a big truck?
2. **How** does impulse explain why boxers like to get hit by bigger and softer gloves?

4.4 Conservation of Momentum and Collisions

LEARNING OBJECTIVES

1. **Describe** conservation of momentum.
2. **Explain** the difference between elastic and inelastic collisions.

Conservation of Momentum

During a free skating session, someone may accidentally skate into you while you are standing in the center of the rink. What happens next depends on how big that someone is and how fast he was going before bumping into you. A small kid would hardly throw you off, but a large hockey player on skates, going at a brisk speed, would most likely send you flying off in the direction of his incoming velocity. It seems that a part of his large momentum was passed on to you. Is the total momentum of both of you conserved after the bump?

To answer this question, consider another situation—that of a bullet fired from a rifle. We know from Newton's Third Law that two equal and opposite forces are in play. The force exerted on the bullet by the rifle will be the same as, but opposite in direction to, the force of the bullet on the rifle. Since the two forces act simultaneously and are equal and opposite, their impulses are also equal and opposite. This means that their changes in momentum are equal and opposite as well. If we add these changes in momentum, they cancel each other. In another words, the bullet and the rifle *as a system* experience no net change in their total momentum. Before firing their total momentum is zero, and after firing it is still zero, even though they are both moving. We have arrived at the **conservation of momentum** law:

During an interaction, the total momentum of an isolated system remains constant.

conservation of momentum During an interaction, the total momentum of an isolated system remains constant.

By a system, we mean a collection of two or more objects. The system is isolated when there is no net force acting on it from outside and the forces that are involved act only between the objects within the system.

The law of conservation of momentum means that if we add all of the momenta now and leave for a while, then upon our return, after we add them again, we will get the same number. Like the conservation of energy law, this is an important physical law with broad applications. **Figure 4.10** applies conservation of momentum to a system consisting of a rifle and a bullet.

Conservation of momentum Figure 4.10
The momenta of the bullet and the rifle are equal and opposite. We can reduce the recoil velocity of the rifle by making it more massive.

Initially, the rifle and bullet are at rest and their total momentum is zero.

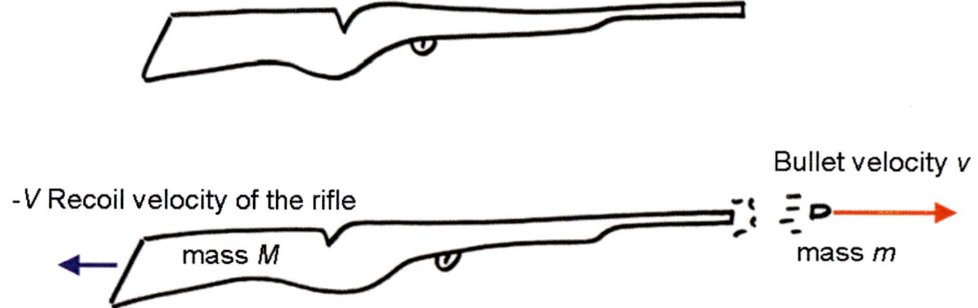

Once the bullet is fired, the total momentum is $mv - MV$. According to the conservation law of momentum, this must equal the initial total momentum, or zero. Therefore, $mv = MV$. Increasing the mass M of the rifle, we decrease its recoil velocity V.

Collisions

Momentum conservation is universal and works during collisions, such as a bump on the ice rink describe above, as long as no outside force is involved. We can also apply it to a collision of two billiard balls coming in contact with each other and then bouncing off one another. Contact forces occur as the two balls strike each other. These action and reaction forces, which act only during the collision, are internal forces. Therefore, the collision of the balls occurs in a closed system and conservation of momentum holds. The loss of momentum of one ball must equal the gain of momentum of the other ball.

The conservation of momentum principle is very useful because we do not need to know the details of the interaction—how large the forces are and how long they act. We just need to know some information about the system *before and after* the interaction. We can complete the missing information by applying this principle. One of the most interesting applications is the study of sub-atomic particle collisions, for instance. Not all collisions are the same. Some collisions can be sticky or **inelastic**, whereas others are bouncy or **elastic**. Most collisions are somewhere in between these extremes. Let's examine two special situations: perfectly elastic and perfectly inelastic collisions.

elastic collision A collision during which colliding objects rebound without lasting deformation and during which the kinetic energy of the colliding objects is conserved

inelastic collision A collision in which the colliding objects become deformed, the colliding objects possibly stick together, and some energy is converted to heat

In a perfectly elastic collision, objects bounce off each other without any deformation or generation of heat. Both momentum and kinetic energy are conserved. In a perfectly inelastic collision, the two objects stick together after the collision. Momentum is conserved but kinetic energy is not. Therefore, the distinction is that in a perfectly elastic collision, no mechanical energy is lost.

In an inelastic collision, objects get deformed, even if temporarily, and this deformation generates heat that is lost to the environment. Some of the kinetic energy is also partially converted into heat. Thus, the total kinetic energy of the colliding objects is smaller after the collision.

We can model a bounce of a basketball from the floor as a collision. A firm or a well-pressurized ball is going to be deformed less during its collision with the floor and thus will lose less energy. As a result, it has more energy to overcome gravitational force—that is, it bounces higher. This describes an elastic collision. A soft basketball does not have much of a bounce at all, and its collision with the floor is inelastic.

A billiard game is a good example of almost perfectly elastic collisions. If one billiard ball strikes another one at rest in a head-on collision, the two balls exchange their velocities. That is, the moving ball becomes stationary and the stationary ball moves with the same speed and direction the other ball had. This exchange preserves the conservation of momentum. A car crashing head-on into another one is often an inelastic collision. The two cars stick and move together in the direction of their total combined initial momentum. A lot of their energy of motion is converted into deforming each other, deforming the passengers, and heat. **What a Physicist Sees** illustrates the elastic collisions in Newton's cradle; and **Example 4.6** shows a quantitative example of an inelastic collision between two roller derby skaters and **Example 4.7** shows how we can calculate the speed of a bullet with the use of a ballistic pendulum.

What a Physicist Sees
Newton's Cradle

The original Newton's cradle of balls was devised over four hundred years ago. Today these cradles are available commercially, usually as executives' toys. The cradle consists of a series of identical metal balls suspended from a frame so that they are just touching each other. Each ball is attached to the frame by two wires of equal length angled away from each other. This restricts the balls' movements to the same vertical plane.

If you pull away one ball and let it fall back, it strikes the first ball in the series and comes to a dead stop. The ball on the opposite side of the series almost instantly swings out in an arc that would be expected of the first ball. If you pull away two balls, two balls on the other side swing out. This is quite puzzling. Why do we always see the same number of balls swing out as the number of balls initially pulled away and released?

To a physicist, this device demonstrates the conservation of momentum and kinetic energy in perfectly elastic collisions, as illustrated in the following diagrams.

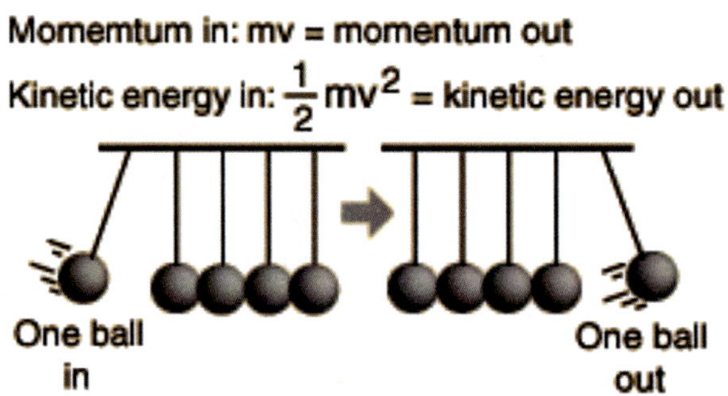

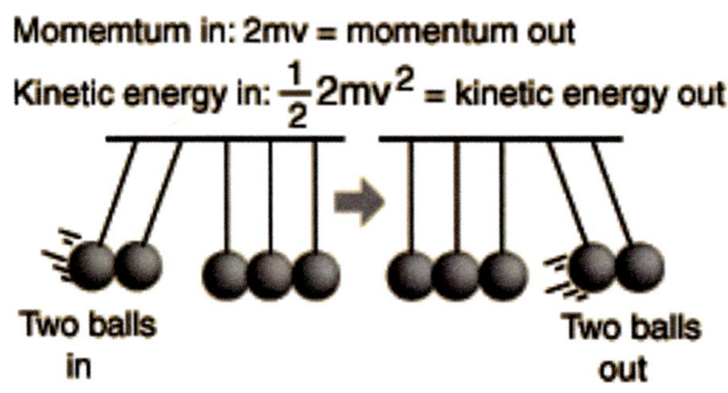

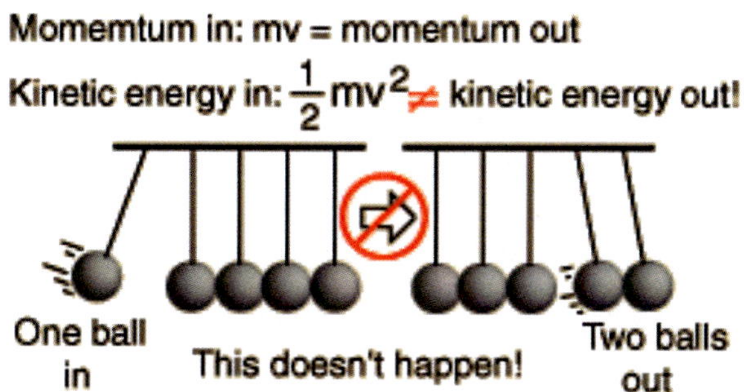

In this case, conservation of momentum would require that the two balls swing out with half the speed.

$$\text{Momentum out} = 2m\frac{v}{2}$$

But this gives

$$\text{Kinetic energy out} = \frac{1}{2}2m\frac{v^2}{4}$$

This would amount to a loss of half the kinetic energy, which cannot happen in elastic collisions.

Think Critically…

What causes some energy dissipation here?

Example 4.6

Inelastic collision between two roller derby skaters

A roller derby skater moving forward with a speed $2v_0$ collides with another skater, also moving forward but only with speed v_0. If both skaters have the same mass m and move together after the collision, what is their final speed?

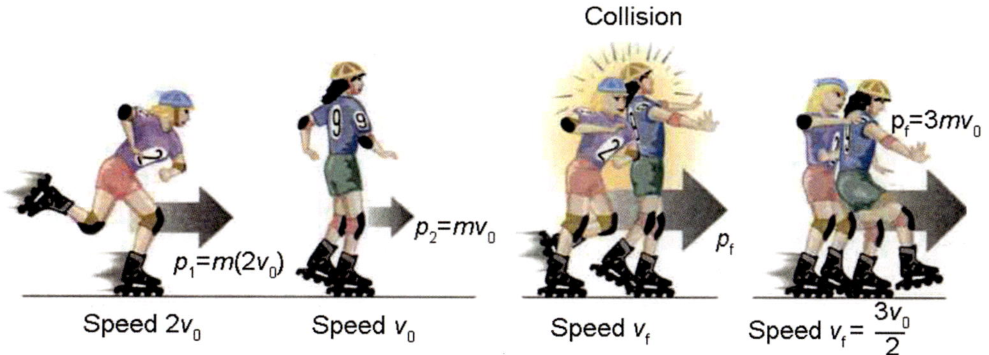

Solution:

The initial momentum p_1 of the first skater is $p_1 = m(2v_0)$.
The initial momentum p_2 of the second skater is $p_2 = mv_0$.
Their total initial momentum is $p_i = p_1 + p_2 = 2mv_0 + mv_0 = 3mv_0$.

Their total final momentum p_f is conserved after the collision. Therefore, their final momentum is $p_f = 3mv_0$. Knowing that final momentum can also be described as $p_f = 2mv_f$, we can solve for v_f:

$$3mv_0 = 2mv_f$$

or

$$v_f = \frac{3v_0}{2}.$$

Example 4.7
Ballistic pendulum
A bullet of mass m hits and imbeds into a large wooden block of mass M, suspended by a string of length L. As a result the block rises to a height h. Find the velocity of the bullet.

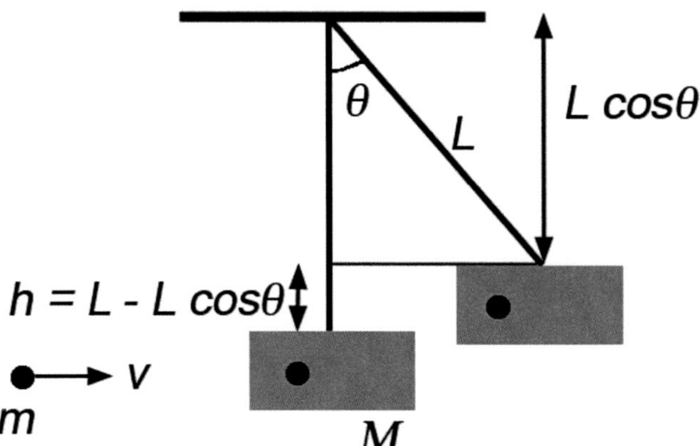

Solution:

Conservation of momentum:

$$mv = (M+m)V$$

where V is the velocity of the block and bullet right after the bullet imbeds itself into the block.

Conservation of energy:

$$1/2\ (M+m)V^2 = (M+m)gh \text{ or } V = \sqrt{2gh}\ .$$

Therefore,

$$v = \frac{(M+m)}{m}V = \frac{(M+m)}{m}\sqrt{2gh}\ .$$

CONCEPT CHECK

1. **What** principle explains rocket thrust?
2. **What** type of collisions do we get during a cream pie food fight?

Summary

4.1 Work, Energy, and Power

- **Work** is the product of a force F in the direction of motion and the distance moved d.
  Work = Force × distance
- A moving object has energy of motion that we call **kinetic energy**. The kinetic energy of the object depends on its mass and speed. We define it as

$$\text{Kinetic energy} = \frac{1}{2}\text{mass} \times \text{speed}^2\ .$$

- The gravitational **potential energy** of an object is calculated by multiplying its mass *m* by the acceleration due to gravity *g* and its height *h*.
  Gravitational potential energy = mass × acceleration due to gravity × height
- **Power** is the rate of using energy and is equal to the amount of work done per unit time, or

$$\text{Power} = \frac{\text{Work}}{\text{Time}}$$

4.2 Energy Transformations

- Work done on an object equals the change of the kinetic energy of the object.

$$\text{Work} = \Delta\text{KE}$$

  We call this relation the **work-energy theorem**.
- Energy cannot be created or destroyed; it may be transformed into another form, but the total energy never changes.

4.3 Momentum and Impulse

- **Momentum** is the product of mass and velocity:

$$\text{Momentum} = \text{mass} \times \text{velocity}$$

- **Impulse** is the product of net force acting on an object and the time during which it acts. Impulse equals change of momentum.

4.4 Conservation of Momentum and Collisions

- The total momentum of an isolated system remains constant.
- **In elastic collisions,** colliding objects rebound without lasting deformation and the kinetic energy of the colliding objects is conserved.
- **In inelastic collisions,** the colliding objects become deformed and possibly stick together, while some energy is converted to heat.

Key Terms

work
energy
potential energy
kinetic energy
work-energy theorem
conservation of energy
power
momentum
impulse
impulse-momentum relation
conservation of momentum
elastic collision
inelastic collision

Critical and Creative Thinking Questions

1. You throw a baseball straight up. Describe its energy transformation as it moves up and then returns back to your hand.
2. You have more than one force acting on an object that moves a distance *d* as a result. How do you find work done on the object?
3. A slow-moving car may have more kinetic energy than a fast-moving motorcycle. How is that possible?
4. What is the difference between energy and power?
5. Which has a greater momentum—a flying mosquito or a heavy truck at rest?
6. What would happen to two astronauts floating in free space if they were to play catch with a baseball?

7. How can you increase impulse?
8. Explain the principle of air bags in cars.
9. If you drop a glass on concrete, it may break. On the other hand, if you drop it on a carpet, it may stay intact. Why?
10. You hit a hard volleyball with your hand. It hurts. Then you let some air out. Hitting it now feels much nicer. What physics principle explains this?
11. Two identical objects moving in opposite directions collide with each other head on. They stick together. If their speeds are identical, will they be moving to the left, to the right, or not at all after the collision?
12. A billiard ball collides in an elastic head-on collision with a second stationary, identical ball. What happens to the first ball after the collision?
13. What are sports examples of transfer and conservation of momentum?
14. If a book falls off a table and crashes to the floor, is work done?
15. Mary and Jane are strong twins. Who does more work: Mary who lifts one block of 100 N over a certain distance at a certain speed, or Jane who lifts 10 blocks of 10 N over that same distance at the same speed (one at a time)? Note: For these simplified examples of work, we're not taking into consideration any complicating factors, such as air resistance in moving one's body or when lifting items.

Exercises

1. A motorboat exerts a constant force of 300 N on a water skier moving at constant speed. How much work does the boat do if it moves a distance of 1 km?
2. The cliff divers in Acapulco jump off a 28 m cliff above the ocean. If the air resistance is negligible, how fast are they going before they hit the water?
3. How much gravitational energy does a 75-kg mountain climber gain when walking from sea level to a 4,500 m peak?
4. A bobsled team drops a vertical distance of 200 m on a racecourse. If all their initial potential energy were converted to their kinetic energy, how fast would they be going at the bottom? Why is this speed too great?
5. What power does a weightlifter need to lift 200 kg a distance of 1.3 m in 0.7 s?
6. If a hair dryer's power is rated at 1000 W, how much energy does it use in 3 minutes?
7. While fishing, you are standing at one end of your small boat, which is stationary on a pond. You walk to the other end of the boat to get a worm. Your speed is 1 m/s, your mass is 70 kg, and the boat's mass is 10 kg. What happens to the boat? If it is moving, what is its speed?
8. Two natives sit in a canoe with a mass of 30 kg at rest. Suddenly an enemy appears and they dive out simultaneously in opposite directions, one from each end of the canoe. The horizontal velocity component of each native is 4 m/s (relative to water). The mass of each native is 75 kg and 90 kg, respectively. Calculate the magnitude and direction of the initial velocity of the canoe after the natives "leave."
9. A tennis ball with a mass of 0.2 kg arrives at a wall with a horizontal speed of 14 m/s. It rebounds horizontally from the wall with a speed of 9 m/s. It is in contact with the wall 0.15 s. What is the magnitude of the average force of the wall on the ball?
10. A 6-g bullet, traveling at 100 m/s, strikes a bulletproof vest. It comes to rest in about 0.0007 s. What average force will it impart to the fellow wearing the bulletproof vest?
11. A rocket engine blasts out 1500 kg of exhaust gas at an average speed of 2000 m/s every second. Calculate the resulting average thrust developed by the engine.
12. Two identical train cars with a mass of 15,000 kg are traveling toward each other on the same track. One train car has a speed of 5 m/s, going to the right, while the other has a speed of 3 m/s, going to the left. When they collide, they couple. What will be the direction and the final speed of the pair?
13. A 6-kg bowling ball collides head-on with a 2-kg bowling pin. The pin flies forward with a speed of 4 m/s. If the ball continues forward with a speed of 2 m/s, what was the initial speed of the ball?

Chapter 5
Rotational Motion

So far we have studied the *translational* motion of objects, motion between any two points. However, many objects, such as a frisbee, in addition to moving from one point to another, can also rotate. In this chapter we will examine *rotational* motion.

Rotation is part of our daily lives, even more than we might think. We are rotating this very moment because we live on the planet Earth, which rotates about its axis while it revolves around the Sun. Rotation was involved in the creation of the solar system, too. On a practical level, most modes of transportation would not be possible without rotating wheels of some kind. Rotation also stabilizes the motion of many moving objects, such as a bicycle or a football. We even pay to experience rotation in amusement park rides, as shown here. Rotational motion is similar to translational motion in many ways, but as we will see, there are also some important differences between the two.

5.1 Describing Rotation

LEARNING OBJECTIVES

1. **Explain** the difference between rotational speed and linear speed.
2. **Describe** rotational inertia.

Any general motion in space can be described as the sum of two different motions: *translation* and *rotation*. Imagine a frisbee moving through the air (**Figure 5.1**). The motion of the frisbee consists of translation and rotation. We already know how to describe translational motion from Chapters 1 and 2. Here, we will focus on rotation.

Translation and rotation of a frisbee Figure 5.1

As the frisbee moves through the air, all points on the disk are moving through space and experience translational motion. At the same time, all points on the disk (other than its very center) are rotating about its center.

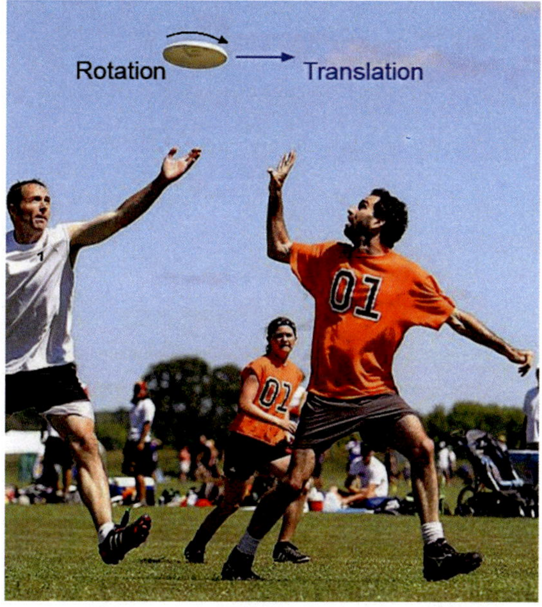

Rotational Speed and Acceleration

Recall that linear speed is the distance traveled per unit of time. A point on the outside edge of a frisbee travels a greater distance during each revolution than a point closer to the center. Therefore, the point on the outside edge must move faster. Recalling that the linear speed of something moving along a circular path is also called tangential speed (see Section 3.1), we can conclude that the tangential speed of any point on the outer edge of the frisbee is greater than that of any point closer to its center.

The standard (SI) unit of distance measurement is a meter. Degrees, radians, or number of rotations are used to measure rotation. Going around a circle once sweeps an angle of 360 degrees. One rotation is defined as 2π radians. See **Figure 5.2.**

Definition of radian – angular unit of rotation Figure 5.2

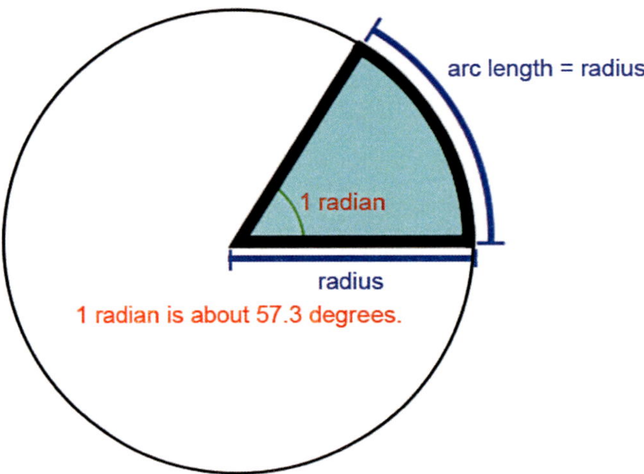

One complete rotation is equivalent to 360 degrees or 2π (about 6.28) radians.

Angular rotation divided by the time of the rotation gives us the **rotational speed**, also known as angular speed. We usually indicate rotational speed with the lower-case Greek letter omega, ω. Possible units of rotational speed include degrees per second, radians per second, or rotations per minute.

rotational speed Angular rotation as measured by the number of rotations, degrees, or radians, divided by the time interval.

Note the difference between rotational speed and tangential speed. Imagine a phonograph record. The entire record rotates at the same rotational speed. In other words, every point on a rigid rotating body has the same rotational speed, regardless of its distance from the axis of rotation. However, the tangential speed depends on the distance of the point from the axis of rotation. The larger the radius, the larger the path taken by the point during each rotation. Since each point on the record rotates during the same time period, any point twice as far from the rotational axis travels twice as far and its tangential velocity is doubled (**Figure 5.3).**

Rotational and tangential speeds of a point on a phonograph record Figure 5.3

a. A point on the outer edge of the record travels twice as far as a point positioned half the radius from the center. The outer-edge point travels twice as fast as the point at half the radius.

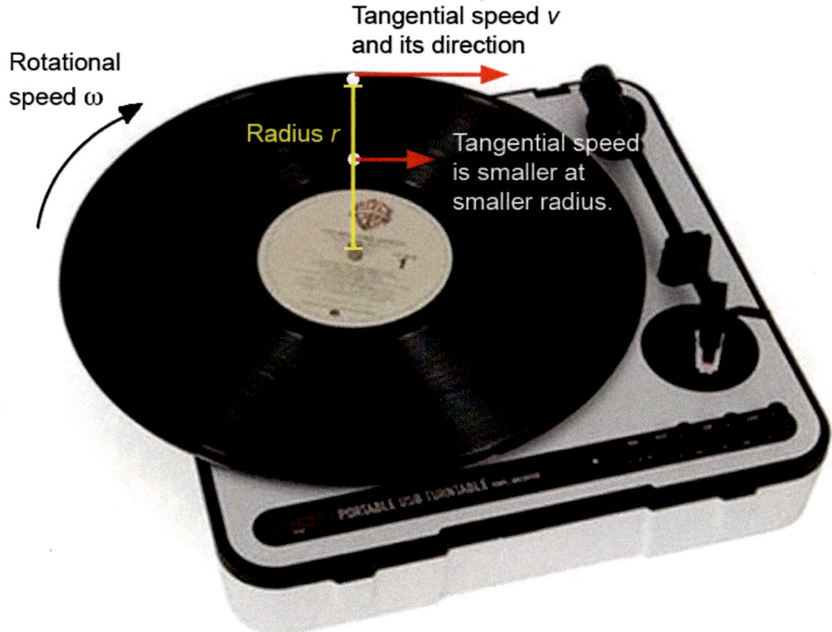

b. Any point on a rotating rigid object has the same rotational speed but a different tangential speed that is proportional to the radius.

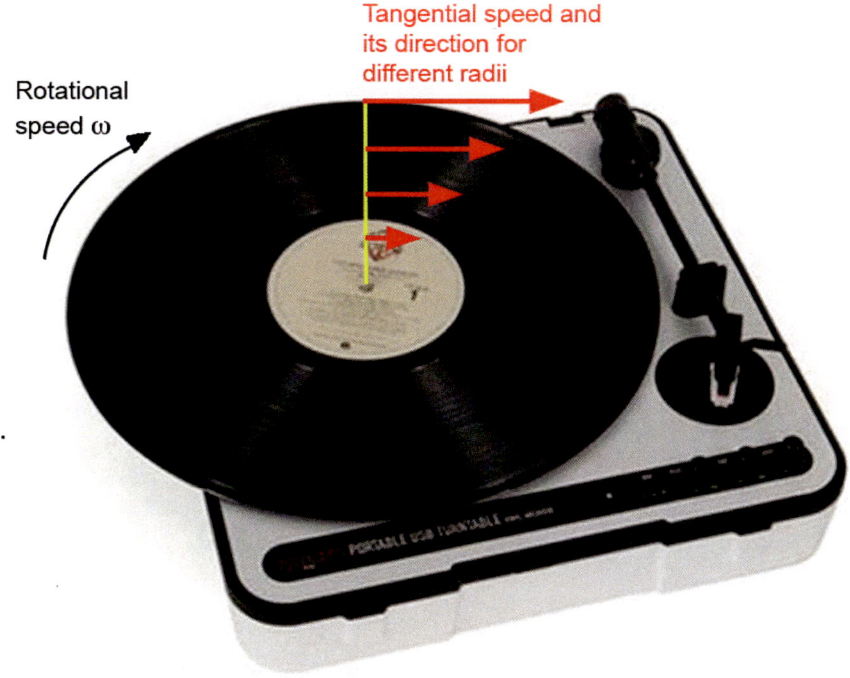

It follows that tangential speed is proportional to both radial distance and rotational speed. We can say that

Tangential speed = radial distance × rotational speed.

In symbols,

$$v = r\omega$$

where v is tangential speed, ω is rotational speed, and r is the radial distance.

Rotating Earth provides a good example. A person standing at the Equator is moving much faster than another person standing at the Arctic Circle, even though their rotational speeds are the same. See **Example 5.1**.

Example 5.1
Rotational speeds on Earth

Earth rotates counterclockwise (as seen from above the North Pole) with a significant speed. Find a) the tangential speed of a person on the equator if the radius of the Earth at the equator is 6,378 km, and b) the distance the person travels in one hour as a result of Earth's rotation.

Solution:

The arrows in the figure denote the tangential (linear) speeds at the equator and the Arctic Circle.

a)

Tangential speed = radial distance × rotational speed.

In symbols,

$$v = r\omega$$

We need to determine the rotational speed first. Earth completes a full revolution or 2π radians in 24 hours. One hour has 3600 seconds (60 minutes × 60 seconds/minute). Therefore, one day has 86,400 seconds (24 hours × 3,600 seconds/hour). Also, 6,378 km = 6,378,000 m. Therefore,

$$\text{Rotational speed} = \frac{\text{angular distance}}{\text{time}} = \frac{2\pi}{86{,}400\text{ s}} = 0.0000727\ \frac{\text{rad}}{\text{s}}$$

Tangential speed = (6,378,000 m)(0.0000727 rad/s) = 463.6 m/s.

b)

In one hour, a person on the equator rotates through a distance of

$$\text{Distance} = \left(3600\frac{\text{s}}{\text{hr}}\right)\left(463.6\frac{\text{m}}{\text{s}}\right) = 1{,}668{,}960\text{ m} \simeq 1{,}669\text{ km} = 1{,}037\text{ miles}$$

A person at the equator moves with a velocity of over 1,000 mph toward the east and is moving much faster than a person standing at the Arctic Circle, even though their rotational speeds are the same.

Engineers take Earth's rotation into account when they launch rockets into orbit. By launching a rocket along the direction of rotation, they need to give it a speed of only 17,000 mph instead of the calculated 18,000 mph that we learned about in Chapter 3, because the rotation of Earth gives it the extra 1000-mph boost. NASA's launching facility is in Cape Canaveral, Florida because Florida is closer to the equator than any other U.S. state, so the rocket has the largest tangential speed. (Another advantage is that firing a rocket into the east from Florida avoids sending it over populated areas.)

So far we have examined uniform circular motion, where the rotational speed was constant. It is an example of centripetal acceleration pointing toward the center of the circle. However, when rotational speed changes, tangential speed changes as well, resulting in *rotational* and *tangential acceleration.* The difference between the two is that rotational acceleration is angular and is measured in angular units, such as radians/s^2, while tangential acceleration is linear and is measured in our regular units, such as m/s^2.

There are many similarities between kinetics from Chapter 1 and rotational quantities when you examine the quantitative relationships. We summarize them all here:

Angular displacement, $\Delta\theta$

$$2\pi \text{ radians} = 360 \text{ degrees}$$

Angular average speed, ω

$$\omega = \Delta\theta/\Delta t$$

t is time.

Angular average acceleration, α

$$\alpha = \Delta\omega/\Delta t$$

Angular kinetics

$$\theta = \omega_i t + 1/2\ \alpha t^2$$

$$\omega_f = \omega_i + \alpha t$$

$$\omega_f^2 = \omega_i^2 + 2\alpha\theta$$

Centripetal acceleration

$$a_c = v_t^2/r$$

where v_t is the tangential speed and r is the radius of rotation.

Tangential velocity v_t

$$v_t = r\omega$$

Tangential acceleration *a*

$$a = r\alpha.$$

Rotational Inertia

The resistance to a translational motion, called inertia, is represented by the mass of an object. Mass also plays a role in determining the resistance to rotational motion, but here we also have to take into account the distribution of the mass relative to the axis of rotation. Imagine having two dumbbells, a short one and a long one. Assume that the mass of the rod connecting the two weights is small compared to the weights. If you hold the short dumbbell in the center, it is easier to rotate than the long one because mass is distributed closer to the axis of rotation.

Rotational inertia, or moment of inertia, gives us an idea of how easily a given body is able to rotate about a given axis of rotation. Larger rotational inertia indicates a larger resistance to rotation, just as a larger mass indicates a larger resistance to translation.

rotational inertia Property of an object that measures its resistance to rotation about a specific axis.

When an object's mass (considered as an equivalent point mass) is concentrated at some radius r from its axis of rotation, its rotational inertia I is simply defined:

Rotational inertia I of a point mass m moving along a circle of radius r
= mass × square of radius

In symbols,

$$I = mr^2$$

The axis of rotation is perpendicular to the plane of the circle and passes through the circle's center. See **Figure 5.4**.

Example 5.2 calculates the rotational inertia of a dumbbell, and **Figure 5.5** shows how rotational inertia can be invoked to explain why a child walks on the rail of a railroad track with extended arms.

Rotational inertia of a rotating point mass Figure 5.4

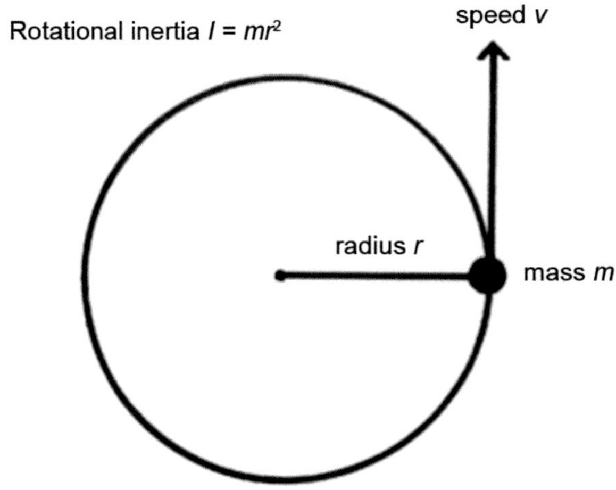

Example 5.2
Rotational inertias of two different dumbbells
Assume that a very light, adjustable titanium rod holds two 10-kg masses in a dumbbell. What is the rotational inertia of the dumbbell about an axis through its center and perpendicular to the rod when the separation of the two 10-kg masses is 0.3 m? What happens if we increase the separation to 0.6 m? (Assume the rod's mass can be neglected.)

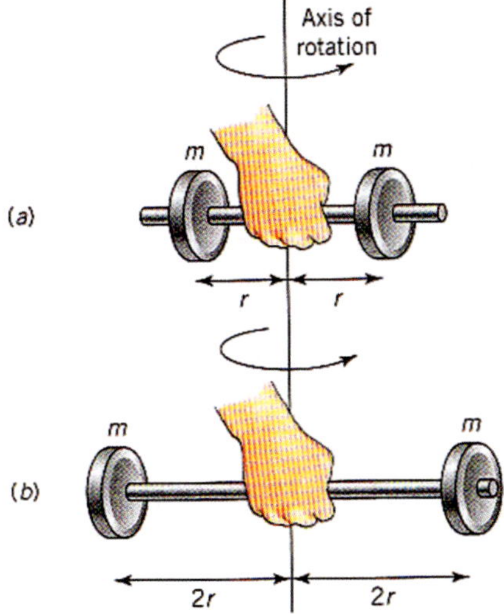

Solution:

We treat the two masses as point masses. Recall our definition of rotational inertia:

Rotational inertia I of a point mass m moving along a circle of radius r = mass m × square of radius r or

$$I = mr^2$$

Each point mass has its own contribution to the rotational inertia. The separation of 0.3 m means that r = 0.15 m, so

$$I = (10\text{ kg})(0.15\text{ m})^2 + (10\text{ kg})(0.15\text{ m})^2 = 0.45\text{ kg-m}^2$$

When we double the separation of the masses to 0.6 m, we get a larger rotational inertia:

$$I = (10\text{ kg})(0.3\text{ m})^2 + (10\text{ kg})(0.3\text{ m})^2 = 1.8\text{ kg-m}^2$$

Doubling the separation increases the rotational inertia by a factor of four.

Rotational inertia and balance Figure 5.5

When a boy walks along a narrow rail, his extended arms increase the distance from his hands and arms to the axis of rotation, which runs through the center of his body. If you think of the arms as a large collection of point masses, then their distances to the axis of rotation increase and his rotational inertia increases as well. Now it is harder for him to tip over.

We list the rotational inertias of different objects about selected axes of rotation. The rotational inertia formulas were obtained by modeling each object as the sum of a large number of point masses. Note that the same object can have a different rotational inertia if you rotate it about a different axis of rotation.

Rotational inertia of a thin ring of mass M and radius R about an axis through its center

$$I = MR^2.$$

Rotational inertia of a circular cylinder of mass M and radius R about an axis through its center and perpendicular to the plane of the disk

$$I = 1/2\ MR^2.$$

Rotational inertia of a sphere of mass M and radius R about any axis through its center

$$I = 2/5\ MR^2.$$

Rotational inertia of a thin rod of length L rotating about its center

$$I = 1/12\ ML^2,$$

and rotating about its end

$$I = 1/3\ ML^2.$$

Notice that just as for the rotation of a point mass, rotational inertia increases with the mass and the square of the dimension (such as radius or length) of the object. The values for rotational inertia confirm, for instance, that a long rod is easier to rotate about an axis that runs through its center and is perpendicular to its length ($I = 1/12\ mL^2$) than about an axis through one of its ends ($I = 1/3\ mL^2$). The reason is that mass is distributed closer to the rotation axis in the first case than in the second case.

You could make another interesting observation by letting different objects roll down an inclined plane and see which one gets to the bottom first. An object with a smaller rotational inertia relative to its own mass has a smaller resistance to rotation and therefore rolls down any incline with a higher acceleration. For instance, a solid sphere ($I/mR^2 = 0.4$) will win the race against a solid cylinder ($I/mR^2 = 0.5$) and a cylindrical shell ($I/mR^2 = 1$). Similarly, a solid sphere will beat a spherical shell.

CONCEPT CHECK
1. How do rotational speed and linear speed vary as a function of radius on a spinning disk?
2. What happens to your rotational inertia when you spread your arms while walking on a railroad track, and why does this stance stabilize you?

5.2 Torque

LEARNING OBJECTIVES

1. **Define** torque.
2. **Describe** all necessary conditions for static equilibrium.
3. **Explain** center of mass.

You already know that to get an object at rest to start moving, you need to apply a force. To rotate an object at rest, such as a wheel or a door, you also need to apply force, but here the point of application of the force is also important. Think of what you do when you try to rotate any object, such as a bike wheel or a door. You instinctively apply force farther away from the axis of rotation because starting the rotation feels easier. It is not just the force, but also its point of application that determines the ease of getting an object to rotate.

Torque

Opening a door, you apply a force on the door handle. This force causes rotation, and the door opens. You have created a **torque**. The door rotates around its axis of rotation, which runs vertically through its hinges. The force and the position of the force relative to the axis of rotation determine the size of the torque (**Figure 5.6**).

Torque on a door Figure 5.6
The child pulls on the door handles, which are farthest away from the axis of rotation (the hinges), to create larger torque. Pulling on the door closer to the axis of rotation would cause a smaller torque, making it harder to open the door. Eventually, pulling on the hinges, it would be impossible to open the door—the torque would be zero.

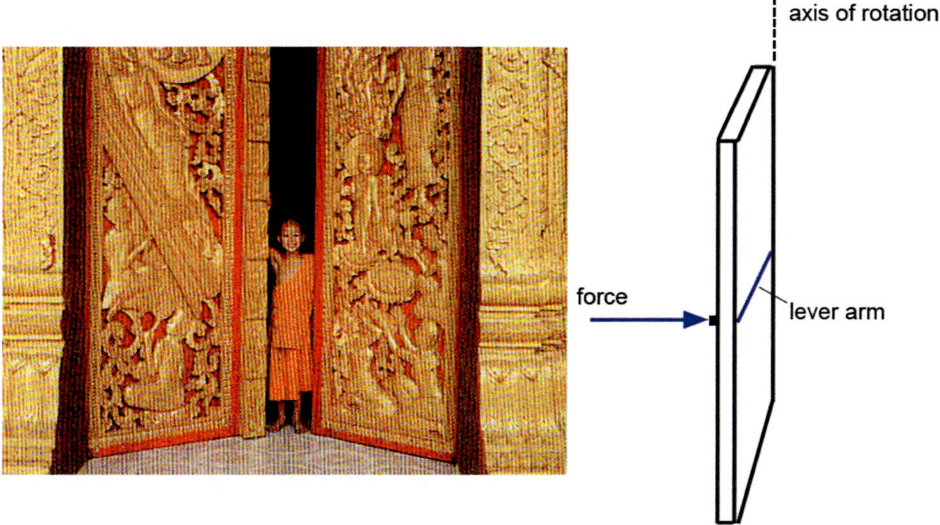

torque Product of a force and the lever arm, which is the shortest distance from the axis of rotation to the line of force. Tendency of a force when applied to a rigid object to cause its rotation.

We define torque τ as the product of the force that tends to produce rotation and a lever arm:

Torque = force × lever arm or

$$\tau = Fr$$

The lever arm is the shortest distance from the axis of rotation to the line of force. This shortest distance is also a perpendicular line from the axis of rotation to the line of force. See **Figure 5.7** for two examples. Metric units of torque are newton-meters.

Common examples of torque Figure 5.7

a. Torque on a bicycle wheel

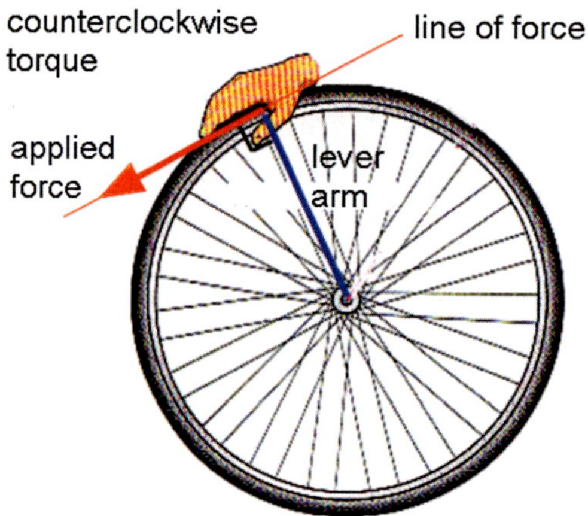

Line of force is perpendicular to the lever arm.

Torque on the wheel = applied force × lever arm

b. Torque on a wrench

Torque = force x lever arm

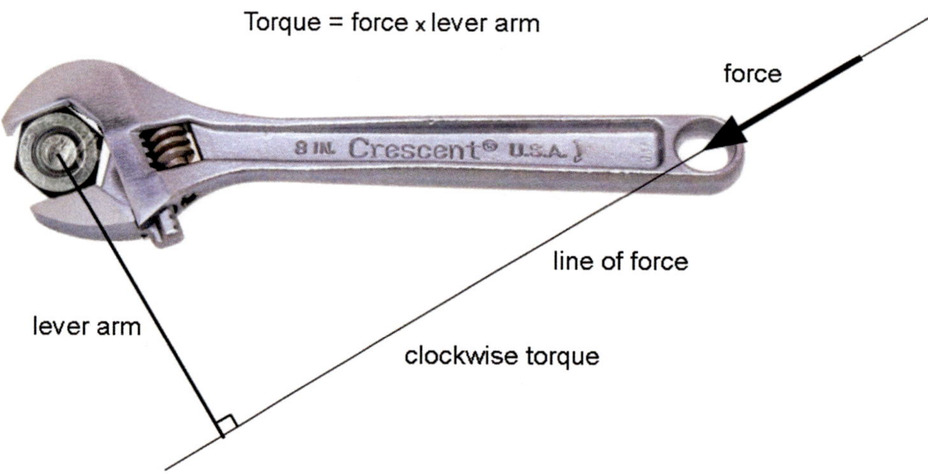

You may remember getting a flat tire and having to change it. Sometimes it is difficult to remove the tightly screwed lug nuts and you need to find a way to generate large torque. Getting a longer wrench would make it easier to loosen them because the longer wrench with a longer lever arm can create a larger torque. If you could not find a longer wrench, you might have tried to put your body weight on it by pushing on it with your foot, though that could be dangerous. Either way, you would have created larger torque. **Example 5.3** shows a numerical example.

Example 5.3
Torque on a lug nut
The torque required to loosen a lug nut on a tire has a magnitude of 80 N·m. What is the minimum force that you need to apply to a wrench to loosen it, using a wrench that has an effective lever arm distance of 20 cm?

Solution:

Recall that

torque = force × lever arm or

$\tau = Fr$

Therefore,

80 N m = F (0.2 m), or F = 400 N.

If you compare this force to weight, the object's mass having this weight would be 40.8 kg. This is not a small force.

We learned in Chapter 3 that when we have several forces acting on an object, we add them to find the net force. Similarly, when we have two or more forces acting on an object and they cause different torques, we sum the corresponding torques to find the net torque. We add clockwise torques separately from counterclockwise torques. Then we assign a positive sign to the net counterclockwise torque and a negative sign to the net clockwise torque. In the final step, we simply add the net counterclockwise and clockwise torques.

Recall Newton's second law $F=ma$. We have a rotational equivalent of this law. We replace force with torque τ, mass with rotational inertia I, and acceleration with angular acceleration α to write

torque = rotational inertia x angular acceleration

$$\tau = I\alpha.$$

In another words, angular acceleration is produced by net torque. If several forces produce individual torques, we find the resultant torque before we can apply this equation.

Center of Mass

Sometimes we need to determine the torque caused by the weight of an extended body. For the purpose of calculating torque, we consider all the weight of the object to act at a single, definite point. This point is called the center of gravity. The center of gravity is same as the center of mass if the entire object is exposed to the same gravitational acceleration, as is typical here on Earth, and we will use the term **center of mass** here.

For a symmetrical object, where the mass is distributed uniformly, the center of mass lies at the geometrical center of the object. For instance, the center of mass of a uniform sphere is its geometrical center as well. Similarly, the center of a frisbee is its center of mass. The center of mass of an irregular object can be determined by hanging it by a string attached to two different points of the object, one at a time. Each time, extend the line of the string by extrapolating it by a pen onto the object. The intersection of the two lines determines the center of mass. We can use the center of mass point to describe translational motion of any extended body, such as a wrench, and also the rotation of the body about its center of mass.

The center of mass of an object does not have to be within the object. For instance, a hoop and an empty spherical shell both have their centers of mass in their geometrical centers where no mass exists. A high jumper clearing a bar has her center of mass stays underneath the bar.

We can calculate the center of mass in one dimension easily. See **Example 5.4**.

Example 5.4
Center of mass of three point masses on a line
Three point masses, m_1, m_2, and m_3, positioned at x_1, x_2, and x_3 along a line. Find their center of mass.

Solution:
The torque created by the center of mass at x_{cm}, where the entire mass $m_1 + m_2 + m_3$ *is located,* is equivalent to the sum of the individual torques by each of the three point masses. Therefore, we get

$$x_{cm} = (m_1x_1 + m_2x_2 + m_3x_3)/(m_1 + m_2 + m_3).$$

Mechanical Equilibrium

Sometimes rotation and translation are not desirable. A worker on a ladder certainly does not want to crash because of a sliding or rotating ladder. We have already learned that the sum of all forces acting on an object must be zero to have zero acceleration. This means that in the absence of a net force, an object initially at rest remains at rest. While this is true, this zero-force sum condition does not exclude the possibility of rotation of the object about a fixed pivot point. **Figure 5.8** shows one such situation.

Two equal and opposite forces can cause rotation Figure 5.8
The sum of the two forces acting on the ruler is zero. Nevertheless, the forces create a torque producing clockwise rotation.

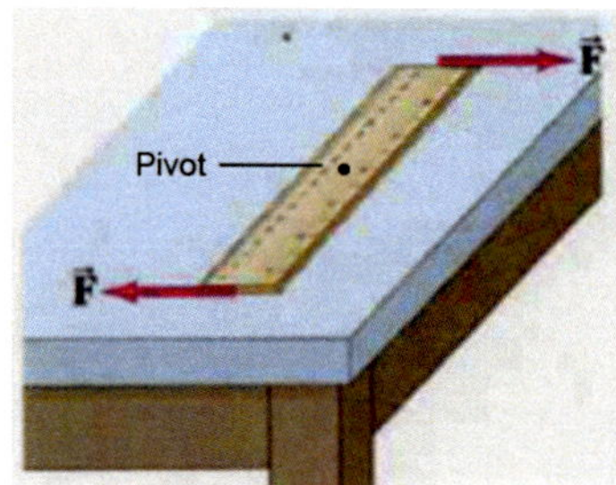

For an object to be in **mechanical equilibrium**, any possible clockwise and counterclockwise torques acting on it must balance and all forces acting on the object must add to zero (**Figure 5.9**). **Example 5.5** shows how a child's seesaw reaches equilibrium.

mechanical equilibrium The state of an object in which all forces acting on the object add to zero and all torques acting on the object add to zero.

A ruler in mechanical equilibrium Figure 5.9
Equilibrium occurs because: Force acting up = sum of forces acting down
Counterclockwise torque = clockwise torque

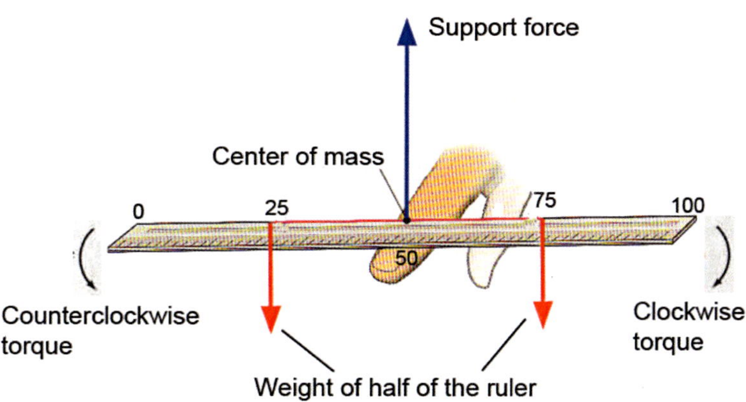

Example 5.5
Seesaw equilibrium
Two children are sitting on a seesaw. The boy, whose mass is 35 kg, sits 2.0 meters from the pivot point. Where must his friend, whose mass is 27 kg, sit to put the seesaw in equilibrium?

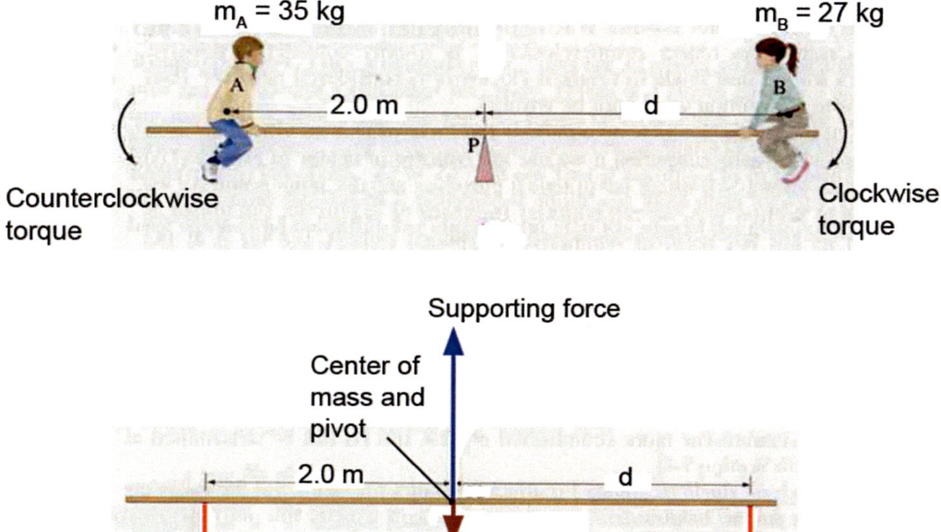

In equilibrium, all forces and all torques add up to zero. Therefore,

Counterclockwise torque by the boy = weight of boy × lever arm =
= (35 kg)(9.8 m/s^2)(2.0 m) = 686 N m

Clockwise torque by his friend = weight of friend × lever arm =
= (27 kg)(9.8 m/s^2) d = 264.6 d N m

The two opposing torques must be equal to have equilibrium:

$$686 = 264.6\ d$$
$$d = 686/264.6 = 2.6 \text{ m}$$

His lighter friend must sit 2.6 m from the pivot center.

CONCEPT CHECK

1. How could you increase torque on a wrench while replacing a flat tire?
2. What is the necessary condition to have a balanced seesaw?
3. How could you find the center of mass of a painter's palette?

5.3 Angular Momentum

LEARNING OBJECTIVES

1. **Define** angular momentum.
2. **Explain** conservation of angular momentum.

We have learned that any moving object has momentum—the product of mass and velocity. We sometimes call it linear momentum. Similarly, any rotating object has momentum due to rotation, which we call angular momentum.

In this section we define the **angular momentum** of a rotating extended body, such as a figure skater, and also of a rotating point mass. We will show that angular momentum is a conserved quantity under certain conditions. This conservation law becomes useful in analyzing and predicting outcomes of rotational motion of a broad range of objects, ranging from frisbees to Olympic divers to the motion of planets around the Sun.

angular momentum Product of an object's rotational velocity and rotational inertia.

Angular Momentum and Its Conservation

Just as any translating object has a linear momentum, any rotating object has an **angular momentum** L that depends on its rotational velocity and rotational inertia:

Angular momentum = rotational inertia I about the axis of rotation × the rotational velocity ω or $L = I\omega$

Notice the similarity to linear momentum, defined as mass × velocity. Just like linear momentum, angular momentum has direction. However, the direction of angular momentum is not the same as the direction of an object's rotation. Instead, angular momentum is determined uniquely by a right-hand rule shown in **Figure 5.10**. In most situations it also lies along the axis of rotation.

Direction of angular momentum Figure 5.10
Curl the fingers of your right hand in the direction of rotation. The thumb points in the direction of the angular momentum.

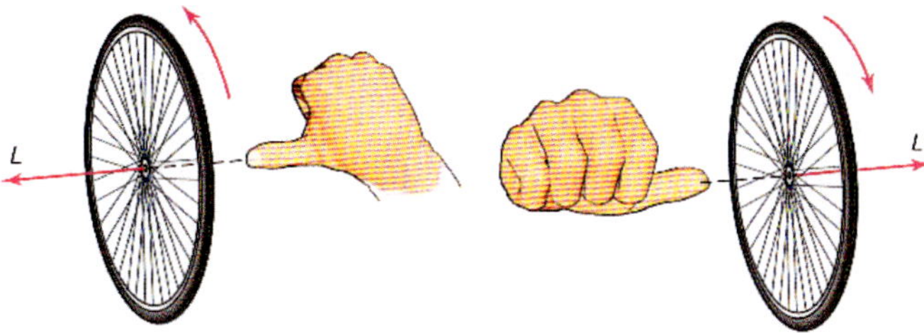

For an object that is small relative to its radius of rotation, angular momentum is simply defined as

> Angular momentum L = object's linear momentum mv × radius of rotation r, or
> $L = mvr$

The motion of a planet around the Sun or the motion of a satellite around Earth could be modeled by this simple expression. See **Figure 5.11.**

Angular momentum of a point mass Figure 5.11
Angular momentum of a point mass is the product of mass *m,* velocity *v*, and radius *r.* It points along the axis of rotation.

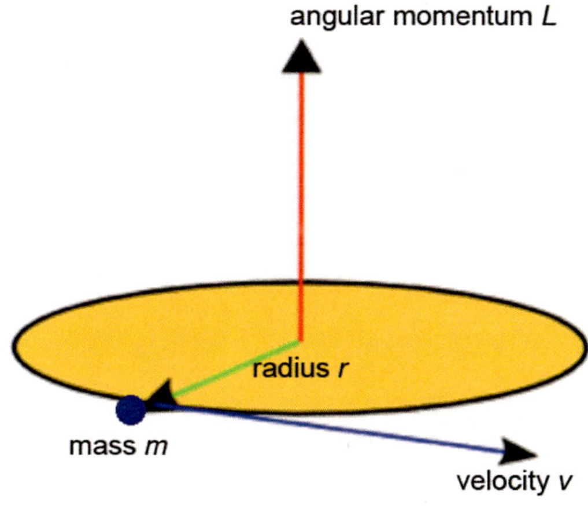

We have already seen that we need a net external force to change the linear momentum of an object. Similarly, we need a net external torque to change an object's angular momentum. Therefore, angular momentum will remain unchanged in the absence of a net torque:

If there is no net external torque present, then the total angular momentum of a system of objects is conserved.

This statement is also known as the law of **conservation of angular momentum.** It is a rotational equivalent of the law of conservation of linear momentum.

conservation of angular momentum When no net external torque acts on an object or system of objects, the total angular momentum remains constant.

The rotation of many figure skaters offers a good example of this conservation law—see **Figure 5.12**

Conservation of a person's angular momentum Figure 5.12

The figure skater starts to spin by stretching out her arms and then bringing them back to her torso as her rotational speed increases. The skater can then break the dizzying spin by throwing her arms back out. (We can assume that the friction between the skater's blades and the ice is negligible, so there is no external, and in this case retarding, torque.) This illustrates how the skater's angular momentum is conserved: by bringing her arms toward her torso, the skater reduces her rotational inertia. Since the product of her rotational inertia and her rotational speed must be the same throughout the spin, the reduced rotational inertia must bring about an increase in her rotational speed.

Angular momentum $L = I\omega$ = constant

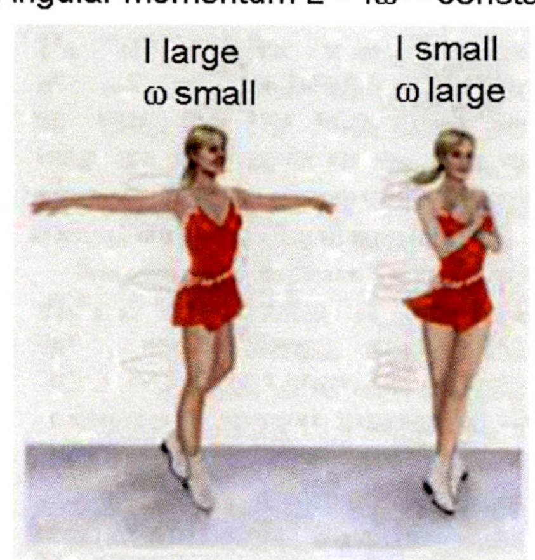

An object does not have to move along a circle to have angular momentum. For instance, all planets in our solar system move in elliptical orbits, and they all have angular momentum. No net external torque acts on the solar system; therefore, the angular momentum of the orbiting and spinning planets remains the same. In practice, this means that if the radius and mass of Earth remain the same, then Earth will continue to rotate with the same angular speed and our night/day cycle will continue to last 24 hours. **Example 5.6** shows how we apply conservation of angular momentum to a satellite orbiting Earth in an elliptical orbit.

Example 5.6

Conservation of angular momentum of a satellite

A satellite is placed in an elliptical orbit around Earth. Its point of closest approach is called *perigee* and is 7,000 km from the center of Earth, while its farthest point is called *apogee* and is 24,000 km from Earth's center. If the speed of the satellite is 8 km/s at perigee, what is it at apogee?

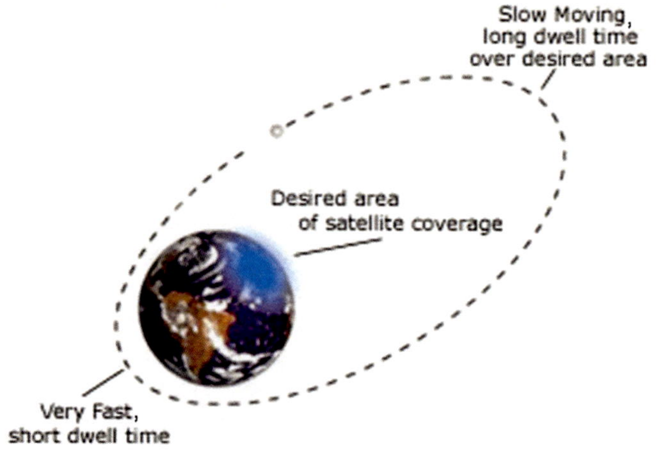

Solution:

No net external torque acts on the satellite, and therefore its angular momentum is conserved.

(mass × speed × radius) at perigee = (mass × speed × radius) at apogee.

Since mass is the same on both sides of the equation, we can divide it. Inserting the numbers, we have:

$$(8 \text{ km/s})(7000 \text{ km}) = (24{,}000 \text{ km})v, \text{ and } v = 2.33 \text{ km/s}.$$

By traveling slowly on the apogee side, the satellite can dwell a long time over the area of the Earth of interest.

Applications of Conservation of Angular Momentum

Conservation of angular momentum means that the magnitude and direction of angular momentum are *both* conserved. For example, if a quarterback threw a football without spin, it would tumble through the air. By putting spin on the ball, the quarterback establishes an angular momentum that stabilizes the football's flight. See **Figure 5.13**.

Direction of angular momentum of a football Figure 5.13
Spinning of the ball establishes its angular momentum whose direction is then fixed in the absence of net external torque, preventing the ball from tumbling. In practice, a small retarding torque exists due to air friction, which makes the direction of the football's angular momentum sweep out a small circle during its flight.

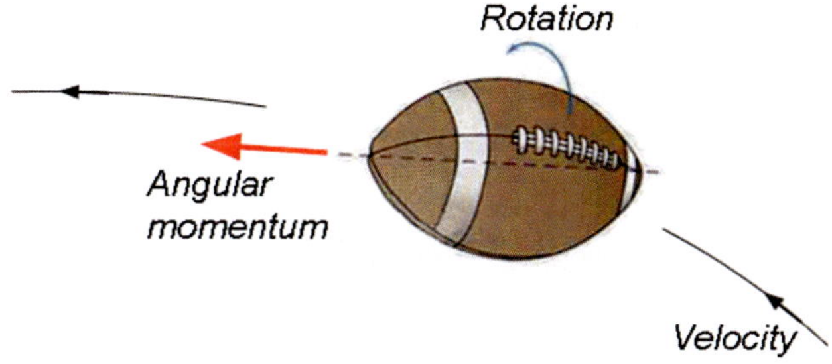

Another example of the conservation of angular momentum is the operation of a helicopter's rotors. If a helicopter had only a single rotor, its whole body would rotate in the direction opposite that of the rotor to conserve angular momentum. That is why large helicopters have two oppositely rotating rotors. See **Figure 5.14**.

A somewhat unexpected example of angular momentum conservation occurs during a motorcycle stunt man's jump maneuver. While in mid-air, a motorcycle stunt man may find himself in an unstable situation where his front wheel drops too far below the motorcycle's center of mass. At different times, the opposite if this may occur when the front wheel loops upward. It is interesting how he compensates by revving up or revving down the engine to add or subtract angular momentum. See **Figure 5.15**.

Angular momentum of a helicopter's rotors Figure 5.14
To prevent rotation, large helicopters have two oppositely rotating rotors. Smaller helicopters place a small rotor on their tail; its rotation creates a counter torque to oppose the rotation of the helicopter's body. Total angular momentum of the helicopter's rotors is zero.

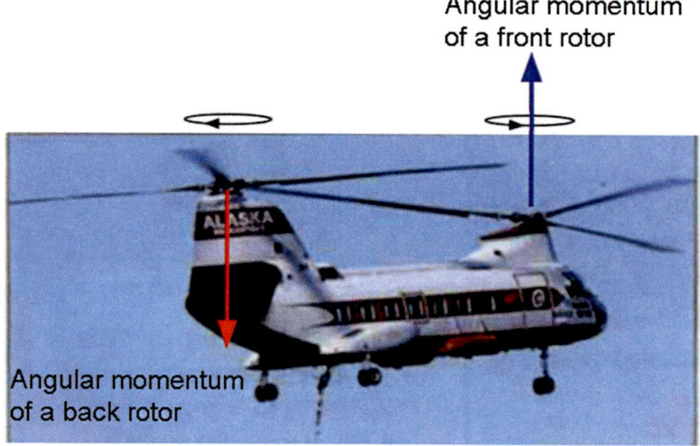

Looping in and looping out by a motorcycle daredevil Figure 5.15
Briefly before landing, it looks as if the rider is about to crash into the ground nose-first. (The front wheel is at a steep downward angle.) Riders call this situation "looping in." An experienced rider can solve this problem by revving the engine while in mid-air. This causes the rear wheel to spin faster and gain angular momentum. The axis of rotation of this additional angular momentum is the rear axle. Because total angular momentum must be conserved, the frame of the motorcycle rotates upward around an axis through the center of mass of the rider and the motorcycle. This gives the needed lift to the front wheel.

The opposite of this situation is called "looping out," during the early phase of his flight, where the front wheel loops upward. The rider can stop the rotation of the rear wheel to bring the motorcycle back to a stable position.

CONCEPT CHECK

1. What is the direction of angular momentum of a second hand on a clock?

2. Why does a figure skater increase her rotational speed by bringing her arms in?

Summary

5.1 Describing Rotation

- Angular rotation divided by the time of rotation gives us the **rotational speed**, also known as angular speed.
- Tangential speed = radial distance × rotational speed.
- **Rotational inertia**, or moment of inertia, describes how easily a given object is able to rotate about a given axis of rotation. Larger rotational inertia indicates a larger resistance to rotation, just as a larger mass indicates a larger resistance to translation.
- Rotational inertia I of a point mass m rotating about a circle of radius r is
$$I = mr^2$$

5.2 Torque

- Torque is the tendency of a force applied to a rigid object to cause that object to rotate. Torque = force × lever arm or
$$\tau = Fr$$
The lever arm is the shortest distance from the axis of rotation to the line of force.
- We can use the **center of mass** point to describe translational motion of any extended body, such as a wrench, and also rotation of the body about its center of mass.
- For an object to be in **mechanical equilibrium**, any possible clockwise and counterclockwise torques acting on it must balance and all forces acting on the object must add to zero.

5.3 Angular momentum

- Any rotating object has an **angular momentum**. The angular momentum can be defined as Angular momentum = rotational inertia I about the axis of rotation × the rotational velocity ω or
$$L = I\omega$$
- **Angular momentum** L of a point mass m moving along a circle of radius r with velocity v is
$$L = mvr$$
- **Conservation of angular momentum:** If no net external torque is present, then the total angular momentum of a system of bodies is conserved.

Key Words

- **rotational speed**
- **rotational inertia**
- **torque**
- **mechanical equilibrium**
- **center of mass**
- **angular momentum**
- **conservation of angular momentum**

Critical and Creative Thinking Questions

1. A rigid object rotates about a fixed axis. Do all points on the object have the same linear speed? If not, which points move the fastest?
2. Does angular speed vary on a rotating rigid object?
3. If two forces produce the same torque, does it mean they have the same magnitude?
4. Why do you extend your arms as you walk on a train rail?

5. Sit-ups are more difficult to do with your hands placed behind your head instead of on your stomach. Why?
6. Why would you use a longer wrench to loosen a stubborn bolt?
7. Can you create a torque by pushing on an object's center of mass?
8. How can you find out if a boiled egg is well done?
9. Why do downhill skiers lean into their turns?
10. Why do quarterbacks throw a football with a spiral motion when they throw a pass?
11. Why do ice skaters bring their arms in once they start to rotate? Would it help to raise their arms above their head? Explain.
12. Why are high divers able to rotate faster when they get into a tuck position?
13. Why is a bicycle more stable when it is moving than when it is stationary?
14. Give one example of conservation of angular momentum.
15. A ball is moving along a circle in the plane of the page. It is moving counterclockwise. What is the direction of its angular momentum?

Exercises

1. What is the equivalent of 1 rev/min in rad/s?
2. What is the angular speed of the minute hand of a clock?
3. What is the rotational inertia of a thin ring about its center if it has a mass of 1 kg and a radius of 1 m?
4. A baton is made of two identical light rods that are attached at their centers and form a cross. The length of each rod is 1 m and a mass of 0.1 kg is attached to each end. Find the rotational inertia of the baton about an axis perpendicular to the plane containing the batons and going through the center of the cross.
5. Two children with masses of 25 and 35 kg are sitting on a balanced seesaw. If the lighter child is sitting 4 m from the center, where is the heavier child sitting?
6. A force of 60 N is applied to a door at different points from the door's axis of rotation, a) $r = 0.80$ m, b) $r = 0.40$ m, and c) $r = 0$. Find the torque in each case.
7. You are holding a 15-kg dumbbell at arm's length. If your arm is 0.8 m long, what torque is the dumbbell exerting on your shoulder?
8. The torque required to loosen a nut holding a flat tire in place on a car has a magnitude of 50 N·m. What is the minimum force that a mechanic needs to apply to loosen it if he is using a 25-cm-long lug wrench?
9. A satellite is placed in an elliptical orbit around the Earth. Its point of closest approach is called *perigee* and is 8,000 km, while its farthest point is called *apogee* and is 25,000 km. Both distances are measured from the center of the Earth. If the speed of the satellite is 8 km/s at perigee, what is it at apogee?
10. If a gymnast performing a forward somersault increases his angular velocity from 3 rad/s to 6 rad/s by tucking in, how does this affect his rotational inertia?

Chapter 6
Fluids

Free divers use only one breath of air during a dive. The sport started with competition among Italian spear fishermen. Raimondo Bucher set the first free-diving world record at 30 m in 1949. By 1960, Bucher's countryman Enzo Majorca reached 49 m, beyond which doctors warned him not to go. However, Majorca went on, breaking his own record at regular intervals until 1974, when he reached 87 meters.

Underwater pressure can be so intense that divers lose feeling in their arms and legs. Lungs shrink to the size of grapefruits, vision blurs, and the heart slows from a normal 70 beats per minute to about 15 beats per minute; blackouts can occur.

Free divers understand the effect of increasing water pressure with depth, and this will also be a focus of this chapter. We will examine the pressure of a liquid and its dependence on depth, and you will see how this affects other properties of liquids, such as buoyancy and flotation. We will apply Newton's laws and other principles of mechanics from earlier chapters to develop these ideas. As you study this chapter, you will discover many similarities between liquids and gases.

6.1 Pressure in Fluids and Pascal's Principle

LEARNING OBJECTIVES

1. **Explain** how pressure changes within fluids at rest.
2. **Describe** Pascal's principle.
3. **Explain** how a hydraulic lift works.

Our lives depend on and are defined by the properties of liquids and gases. For example, the most abundant substance in our bodies is liquid water. The watery environment of our cells promotes chemical processes that nurture life. Without oxygen gas in the air, we could not breathe, and we would die. We will explore the properties of liquids in this chapter and of gases in the next one. However, keep in mind that liquids and gases have many properties in common, such as their ability to flow, which is why we sometimes refer to them both as **fluids**, from a Latin word meaning "to flow." Understanding pressure's behavior within liquids helps us design machines that can lift heavy loads or bind books, for instance, and that improve quality of our lives.

fluids Liquids and gases.

Pressure within Liquids

Recall from Chapter 6 that pressure is defined as an amount of force per area,

$$\text{Pressure} = \frac{\text{Force}}{\text{Area}} \qquad \text{or} \qquad P = \frac{F}{A}$$

What causes the pressure that divers feel on their eardrums as they dive deeper? Try this experiment. Place several books on the top of your head. Does their weight make you uncomfortable? Similarly, the weight of water above the diver creates an uncomfortable pressure. The situation is similar to the pressure caused by the weight of stationary air above your head when you are on land.

To describe the properties of pressure within liquid, we start by looking at still water in an aquarium. Since water is still, there is no motion of any small point-like mass of water. **Figure 6.1** shows the addition of all forces acting on the point and leads us to a conclusion about pressure at an arbitrary point inside a still liquid.

Pressure at any point within a liquid Figure 6.1

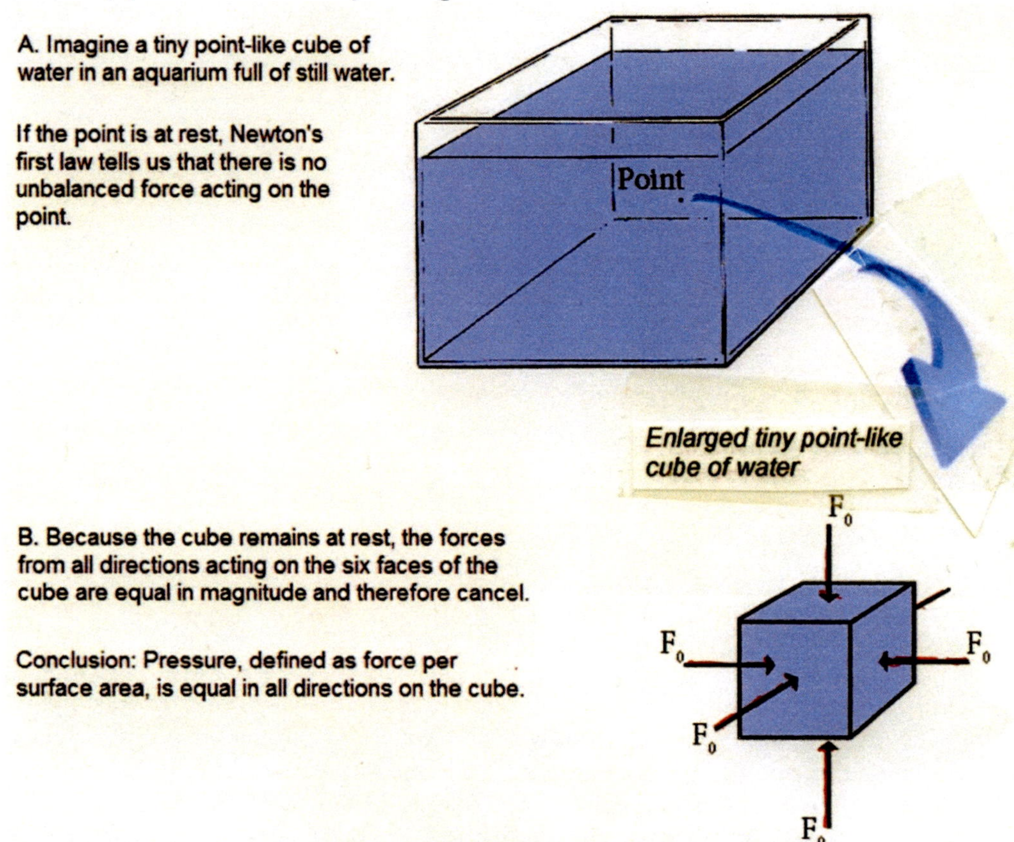

A simple example illustrates another property of pressure in liquids. The shape of a container holding a liquid and the area of its bottom surface do not affect the pressure on the bottom surface. The pressure on the bottom of the container depends only on the height of liquid above the bottom surface. We can perform a simple experiment confirming that pressure at any point at a given depth of connected tubes is the same and also does not depend on the volume of liquid above. We can use this fact to build a simple device that measures vertical level. See **Figure 6.2**.

Let's summarize what we have observed about a liquid at rest so far;

1. At any point inside the liquid, pressure acts equally in all directions.
2. Pressure increases with depth below the surface.
3. Pressure remains the same horizontally across the liquid.
4. Pressure does not depend on the volume of liquid above.

In the next section, we will build on these properties to find Pascal's principle that will lead us to some interesting applications.

Pressure in different vessels Figure 6.2

a. The volumes of the two containers are different, but the pressure on the bottom surface of each is identical because pressure depends only on the height and not the volume of water above the bottom surface.

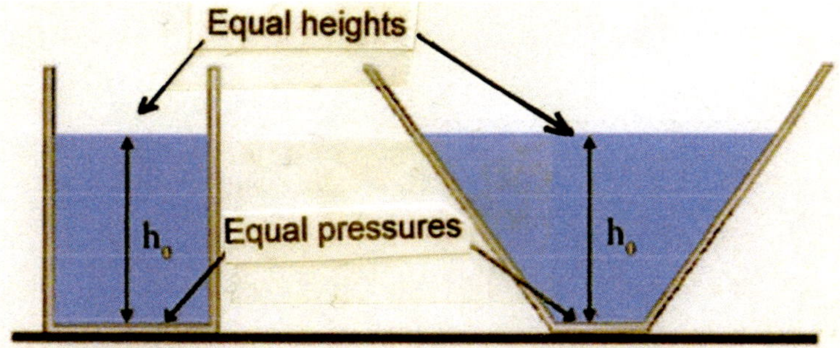

b. Four glass tubes are connected to a common bottom. The points A, B, C, and D are all at the same depth, and the water pressure at these points (P_A, P_B, P_C, and P_D) is the same.

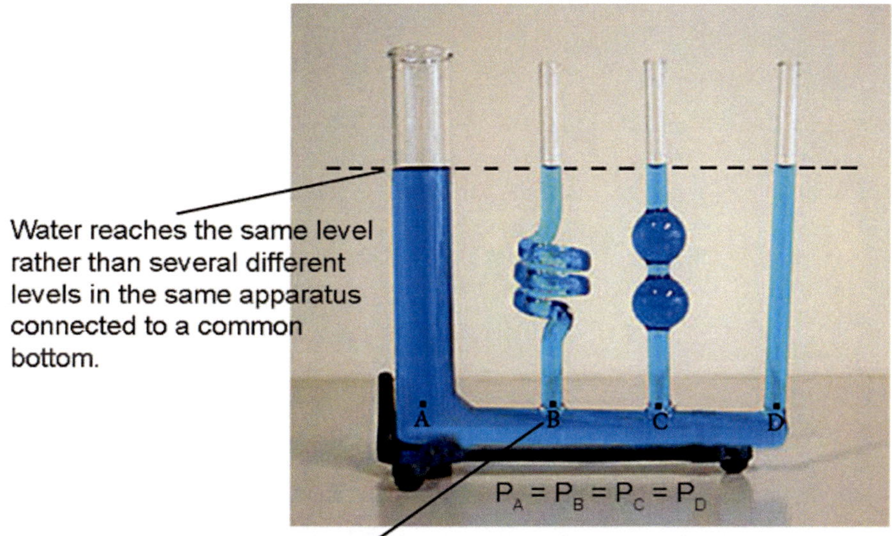

The points A, B, C, and D are all at the same depth, and the water pressure at these points (P_A, P_B, P_C, and P_D) is the same.

If the pressures at the bottoms of the tubes were not equal, the greater pressure at one of the points would force water sideways and then up in one of the tubes until the pressures at the bottom were equal. The experiment confirms that water seeks its own level rather than several different levels in the same apparatus.

c. With this knowledge, you can design a simple device of two water containers connected by a long rubber hose that builders could use to find the same vertical level in different adjacent spaces.

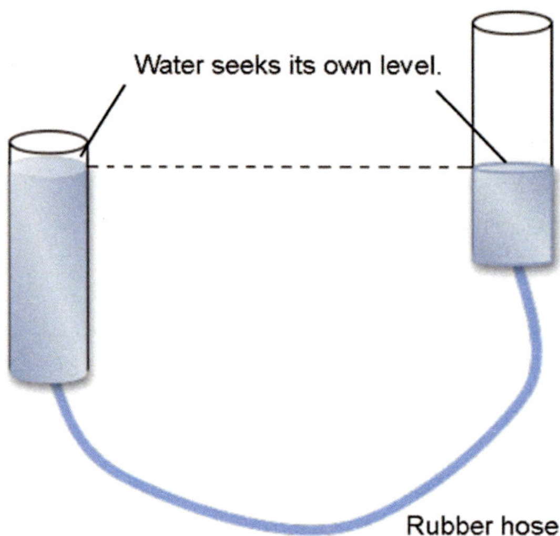

Pascal's Principle and Hydraulics

The properties of liquids at rest become very useful if we enclose a liquid in a cylinder and apply pressure to the liquid with a piston moving inside the cylinder. The pressure is transmitted evenly throughout the liquid. Understanding how pressure is transmitted through liquids enables engineers to design devices that simplify work. For example, hydraulic systems (that is, systems operating with an enclosed liquid) take advantage of the ability of liquids to transmit changes in pressure and multiply forces to create a mechanical advantage. Hydraulic lifts in car repair shops and hydraulic presses used by book binders rely on this property to lift heavy loads and press books, for instance. This mechanical advantage comes about through manipulating how pressure is transmitted through incompressible liquids. The French writer and scientist Blaise Pascal (1623-1662) first discovered this important property and stated it as a general principle.

We can visualize transmission of pressure within liquid by looking at a simple ball model in **Figure 6.3**, where balls represent molecules of a liquid, such as water. This simple model suggests how pressure is transmitted through incompressible liquid. Pascal performed experiments on enclosed liquids and stated **Pascal's principle**, one of the most important and useful principles governing incompressible liquids.

Pascal's principle Pressure applied at any point in an enclosed liquid at rest is transmitted undiminished to all other points within the liquid.

This simple principle is used today in the operation of a large number of practical devices, ranging from the hydraulic lift and hydraulic press to shock absorbers and·airplane landing gear. **Hydraulic lift** typically involves two different-sized pistons in contact with a common fluid reservoir. The fluid is usually mineral oil and transmits pressure from the small piston to the large piston, causing a larger upward force on the large piston (**Figure 6.4**). A hydraulic press works like a hydraulic lift, but the large piston pushes down instead of up.

A model of pressure transmission in liquids Figure 6.3
In the three-ball model of force transmission, a container holds three small balls in contact with one another. The balls represent molecules

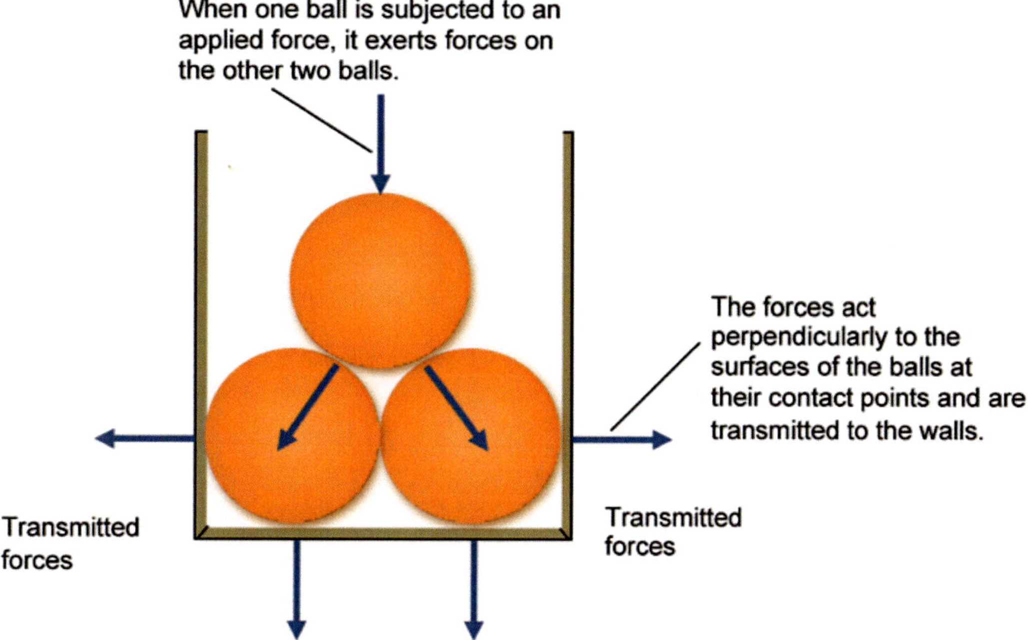

Principle of a hydraulic lift Figure 6.4

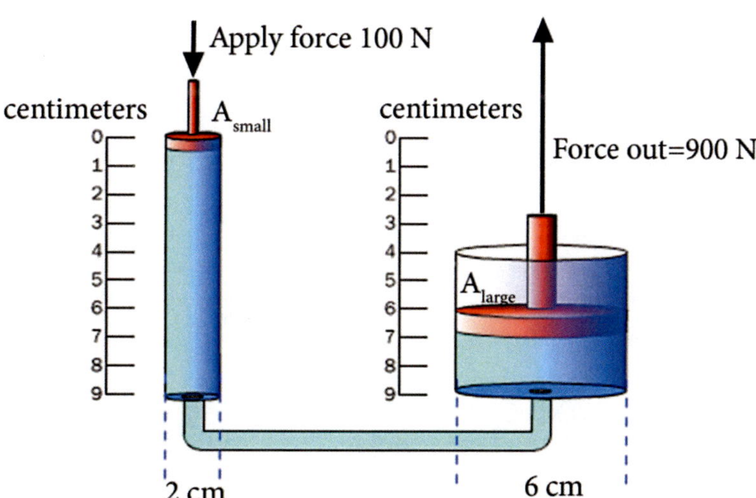

By Pascal's principle, the pressure P acting on the hydraulic oil via the small piston, defined as force F divided by piston area A, or P=F/A, must equal the pressure of the oil acting on the large piston.

Pressure by the small piston on the fluid = Pressure by the large piston on the load

In symbols,

$$\frac{F_{\text{small}}}{A_{\text{small}}} = \frac{F_{\text{large}}}{A_{\text{large}}}$$

By starting with a small force and ending up with a large one, you might think you could generate energy with hydraulic machines. That is not true because the small piston has to move a large distance to move the large piston a small distance. The work the small piston does on the liquid (small force times large distance) equals the work the liquid does on the large piston (large force times small distance). This is another example of conservation of energy.

hydraulic lift A machine in which a small force applied to a small piston is magnified and applied to a large piston to lift a large weight.

Hydraulic machines create a mechanical advantage; they convert a small force into a large force that can move a large object. Or, in the case of a car's brake system, they create a large force to stop a large object. See **Figure 6.5**.

Process Diagram
How hydraulic brakes work Figure 6.5
When you apply a small force to a car's brake pedal, the hydraulic system converts it to a large force acting on the wheel drum.

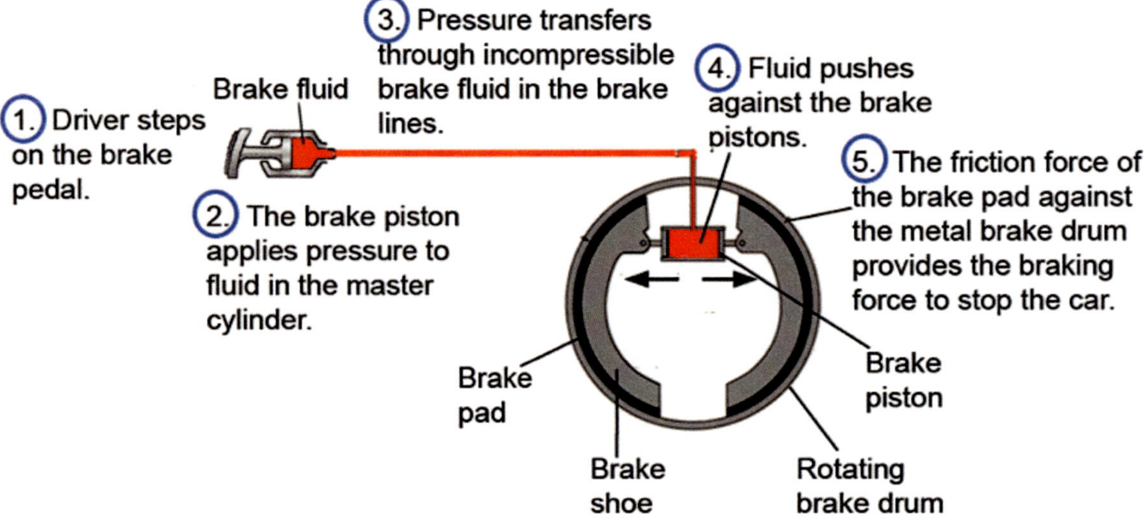

Scuba divers and submarine designers need to understand how pressure increase with depth affects functions of the human body or the structural integrity of a submarine frame. They need to anticipate those pressures. We will go through several steps to determine the total pressure at an arbitrary depth within liquid that is at rest.

Understanding the different pressure measurements in our lives is also important. For example, our doctors report our blood pressure. We check our car's tire pressure regularly. Engineers measure community water towers' pressure in order to regulate the water pressure in our homes. We will show examples of pressure measurements within liquids and also within the atmosphere.

Example 6.1
Lifting a Car with a Hydraulic Jack
A force of 250 N acts on the piston of a small hydraulic jack, having an area of 25 cm^2. The large piston, which supports the load (the output force), has an area of 2500 cm^2. How large a load can the jack lift?

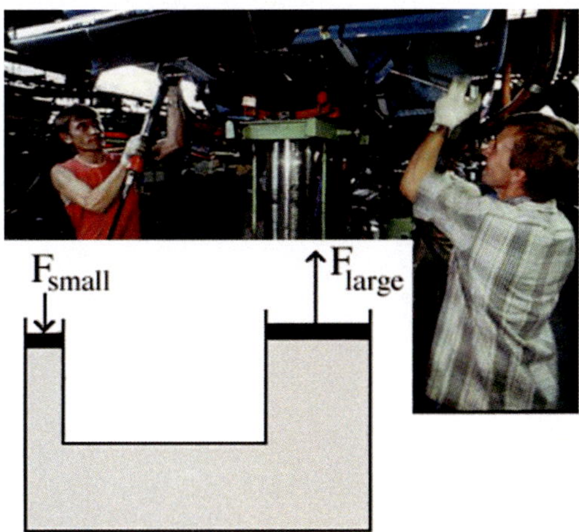

Solution

By Pascal's principle, the pressure P acting on the hydraulic oil via the small piston, defined as force F divided by piston area A, must equal the pressure of the oil acting on the large piston.

Pressure by the small piston on the fluid = Pressure by the large piston on the load

In symbols,

$$\frac{F_{small}}{A_{small}} = \frac{F_{large}}{A_{large}}$$

where A represents the piston area.

$$\frac{250\ \text{N}}{25\ \text{cm}^2} = \frac{F_{large}}{2500\ \text{cm}^2} \qquad \text{and} \qquad \text{Load } (F_{large}) = 25{,}000\ \text{N}.$$

This hydraulic lift creates an output force a hundred times larger than the input force, producing a large mechanical advantage.

Pressure Increase with Depth

We saw earlier in this chapter that pressure below the surface of a liquid depends only on the depth and not on the volume of liquid above. This relationship is true for any fluid – liquid or gas. **Figure 6.6** takes us through several steps to find this **pressure at a given depth**. In the diagram, the liquid is open to the atmosphere, and we take into account the atmospheric pressure as well. (Recall from Chapter 6 the meaning and magnitude of atmospheric pressure, $P_{atm} = 101{,}000\ \text{Pa} = 14.7\ \text{lb/in.}^2$)

pressure at depth h Absolute pressure P_{abs} at depth h = atmospheric pressure P_{atm} + water pressure Dhg due to the weight of water above.

Example 6.2 provides an example of pressure acting on a diver.

The equation for pressure increase with depth Figure 6.6

Imagine an arbitrary point *C* inside a liquid (say, water) at some depth *h*. We want to find the pressure at this point.

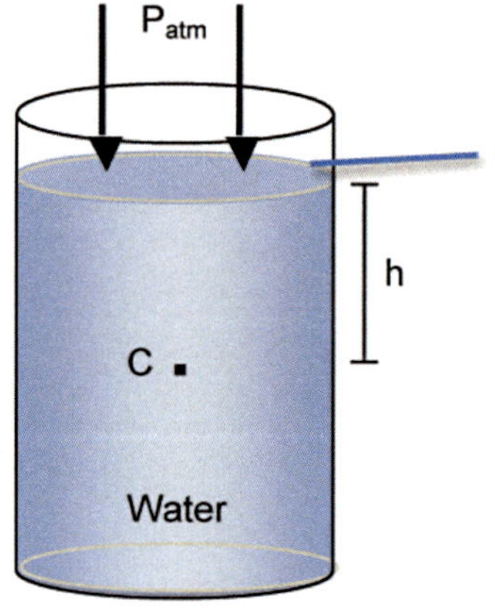

1. Atmospheric pressure P_{atm} acts uniformly on the water's surface. This pressure is transmitted to all points within the water. However, the weight of the water above point C creates additional pressure at this point.

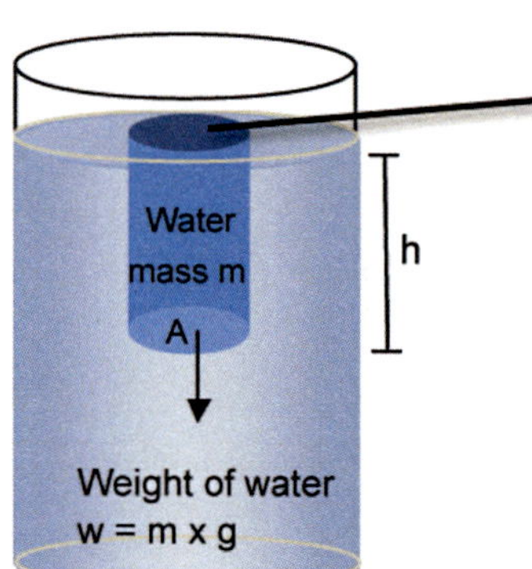

2. To find the pressure at C due to the weight of the water above it, imagine a cylinder of water above C and just below the water's surface. This cylinder has height h and base area A.

   The weight of water within the cylinder is mass m times the acceleration due to gravity, g, and pressure on its bottom surface is

$$\text{Pressure on A} = \frac{\text{Force}}{\text{Area}} = \frac{\text{Weight of water within cylinder}}{\text{Base area of cylinder}}$$

$$P = \frac{m \times g}{A}$$

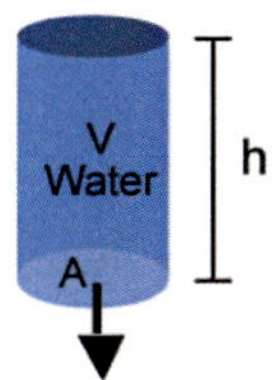

3. We can express the mass of water within the cylinder in terms of its volume, V, and the density of water, D.

$$\text{Density } D = \frac{\text{Mass } m}{\text{Volume } V} \quad \text{or } m = D \times V$$

Weight w is

$$w = m \times g = D \times V \times g$$

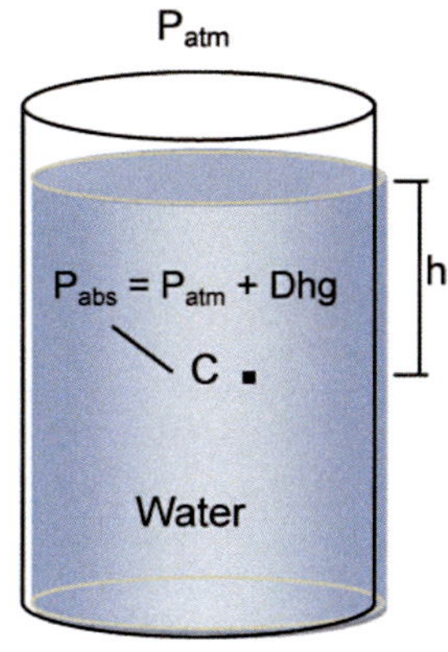

4. Now in the equation for pressure in step 2, we substitute the expression found in step 3 for the weight of water in the cylinder.

$$P = \frac{m \times g}{A} = \frac{D \times V \times g}{A}$$

The volume of the cylinder equals the area of its base *A* times its height h, or V = A x h. So the equation for the pressure becomes

$$P = \frac{D \times A \times h \times g}{A} = Dhg$$

This is the pressure at point C due to the weight of the water above. To find the total pressure at C, or the absolute pressure P_{abs}, add the atmospheric pressure P_{atm}.

$$P_{abs} = P_{atm} + Dhg$$

Example 6.2
Total Pressure on a Diver

Scuba divers easily reach depths of 10 m or more below the surface. Find the total pressure acting on the scuba diver when she is 10 meters under the water's surface.

Solution

Recall that the metric (SI) unit of pressure is N/m^2, also called pascals or Pa. Atmospheric pressure due to the weight of a column of air acts on the water's surface and is transmitted undiminished through the water. Its value is about 101,000 Pa. At the same time, the weight of a column of water 10 m high contributes to the pressure as well. To find the pressure on the diver, simply add the two pressures. This is an application of the above relation $P_{abs} = P_{atm} + Dhg$. The density of water is D = 1000 kg/m^3 (density of seawater is slightly higher). Thus, when the diver is 10 m below the surface of water, the total pressure acting on the diver is

$$P_{abs} = 101{,}000 \text{ Pa} + 1000 \text{ kg/m}^3 \times 10 \text{ m} \times 9.8 \text{ m/s}^2 = 199{,}000 \text{ Pa}.$$

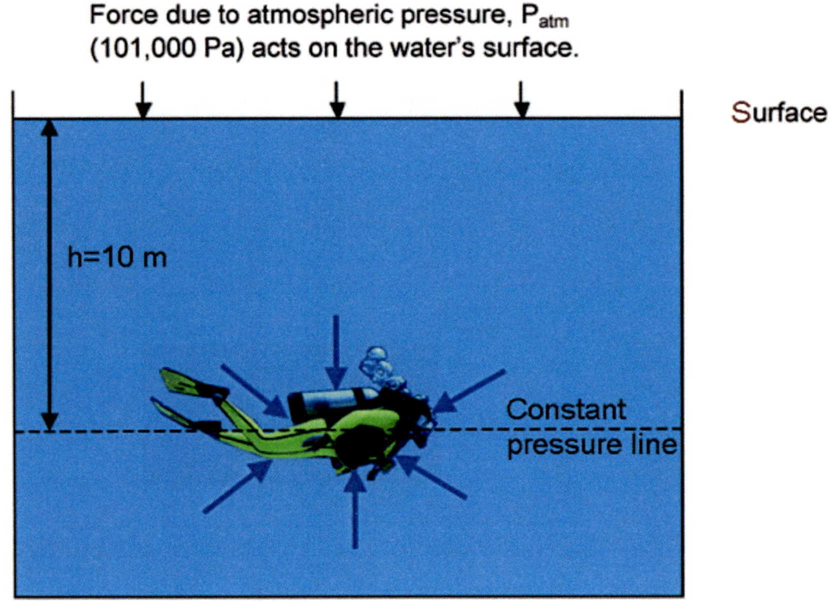

What does this mean? The total pressure at a depth of 10 m is about double what you experience on the surface. This pressure is enough to create painful sensations on your eardrums. At a depth of 100 m, the total pressure is about 11 times what you experience at the surface.

Applications of Pressure in Liquids

If you had a long tube such as a straw, could you control the flow of liquid through it by controlling the pressure difference across the ends? Dip the tube into water, exposing the lower end to atmospheric pressure plus pressure due to the water, and reduce the pressure at the top of the tube by sucking on it. Does the water rise? **What a Physicist Sees** looks at this interesting application.

What a Physicist Sees
Drinking through a straw

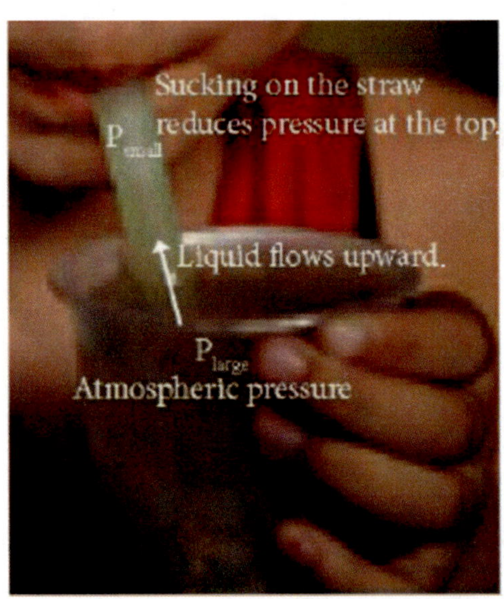

What happens to liquid pressure when the girl in the photo sucks on a straw? A physicist would think first of atmospheric pressure acting on the surface of the drink. This pressure is transmitted through the liquid inside the glass to the bottom of the straw. When the girl sucks on the straw, she reduces the pressure at its top. The difference in pressure between the top and bottom of the straw creates an upward force on the liquid, and it rises.

Think Critically...

If you could create zero pressure at the top of the straw, how high could you raise the water through it? Hint: The water continues rising until the upward force becomes balanced by the increasing downward force of the weight of water in the tube. The pressure due to the weight of a column of water equals the atmospheric pressure at this point.

Another way to put a pressure difference to work is to draw liquid through a syringe, as shown in **Figure 6.7**. **Example 6.3** shows numerically how our knowledge of pressure difference across a U-shaped tube helps us find unknown pressure at one of its ends.

Using pressure difference to draw liquid into a syringe Figure 6.7

As the plunger in a hypodermic syringe is drawn upward, a vacuum or low pressure develops in the cylinder. Normal atmospheric pressure on the surface of the medication liquid in the bottle pushes the liquid up into the syringe.

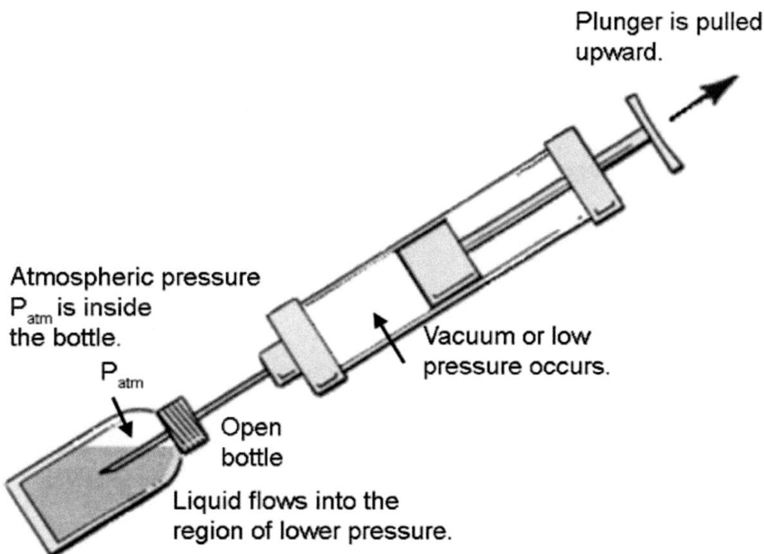

To inject the liquid, you simply insert the needle and depress the plunger. This increases the pressure in the chamber that allows the liquid to flow.

Example 6.3

Measurement of an Unknown Pressure

A U-tube is filled with red water whose ends are exposed to different pressures. As a result, a difference of 10 cm in height occurs between the water levels. If you know the pressure P_1, how can you use this information to find the unknown pressure P_2?

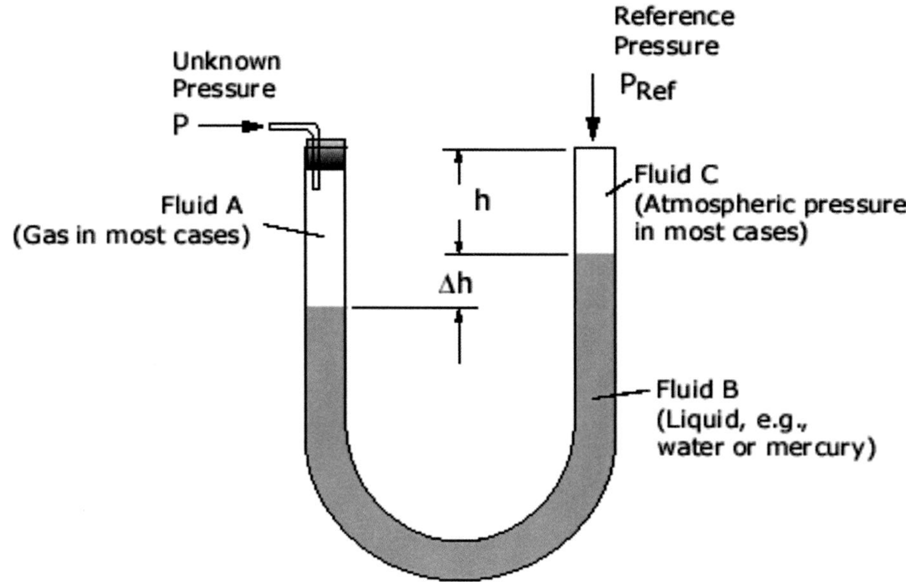

Solution

For the water in the left end of the tube to be lower than in the right end, it must be true that pressure P_2 in the left end of the tube is larger than pressure P_1 in the right end. The positive pressure difference $P_2 - P_1$ causes the water to rise to a height h in the right end.

The weight of the column of water of height h exactly equals and opposes this upward force due to the pressure difference. Therefore,

$$P_2 - P_1 = D \times h \times g = 1000\ \text{kg/m}^3 \times\ 0.1\ \text{m} \times 9.8\ \text{m/s}^2 = 980\ \text{Pa}.$$

We have $P_2 = P_1 + 980$ Pa. For instance, if $P_1 = 101{,}000$ Pa (atmospheric pressure), then $P_2 =$ 101,980 Pa or about 102,000 Pa.

One of the most common applications of pressure measurement is used by weather forecasters and meteorologists every day to measure atmospheric pressure. The instrument they use, called a **barometer,** operates by transferring pressure through a liquid from an open dish to a narrow tube. See **Figure 6.8**. Other types of barometers are also in common use, but they all rely on transferring atmospheric pressure to a measurement scale.

barometer Device that measures atmospheric pressure.

In practice, when talking about pressure, we have to be careful to distinguish between the absolute pressure and gauge pressure. Absolute pressure is the total pressure, including the atmospheric pressure, to which we are all subjected. **Gauge pressure** is equal to the absolute pressure minus the atmospheric pressure. We measure blood pressure as gauge pressure, for instance.

gauge pressure Absolute pressure minus atmospheric pressure.

Pressure transfer through liquid in a barometer Figure 6.8

A tube filled with mercury is inverted into an open dish of mercury. Normal atmospheric pressure acting on the mercury in the dish can support a column of mercury in the tube approximately 760 mm high. As the air pressure varies so does the height of the mercury column.

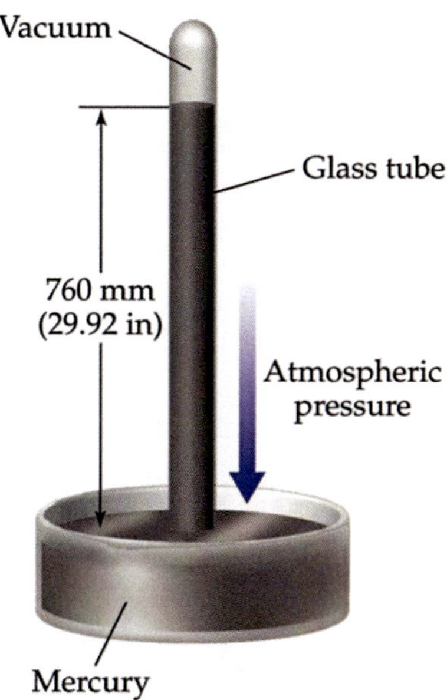

A crucial application of fluid pressure occurs in the human body. Without proper blood pressure in the heart, you could not live because blood would not reach the brain. Contraction of the heart chambers creates pressure that enables blood circulation. The peak heart blood pressure, called systolic pressure, is about 16 kPa (120 mm of mercury) during the contraction phase of a healthy heart. During the relaxation and refilling phase, the minimum, or diastolic pressure, is about 11 kPa (80 mm of mercury). **Figure 6.9** shows the heart's blood circulation and pressure measurement.

Blood pressure in the heart Figure 6.9

The human circulatory system transports about 6 liters of blood through the body every minute.

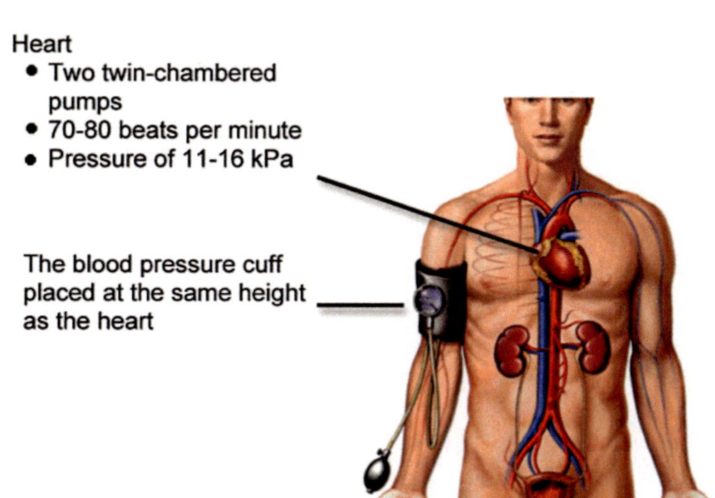

You might experience a change in blood pressure if you stand up too quickly after lying down for a while. Feeling a little dizzy is normal. The reason is that the blood pressure in the brain drops from about 13 kPa (100 mm of mercury, the average of 120 and 80 mm of mercury) to about 9 kPa (65 mm of mercury). Not enough blood or oxygen flows to the brain, so the blood vessels in the head expand to compensate for the pressure drop. This adjustment is not instantaneous, and for a moment dizziness occurs.

When a patient receives intravenous fluid through a needle in a vein, the fluid must overcome the gauge pressure of the vein. To do that, a nurse might raise the pouch with the fluid above the patient's arm, creating external pressure due to the weight of the fluid. (The nurse could also use an IV pump to increase the external pressure.) **Example 6.4** gives a numerical example for administering a solution in a vein.

Example 6.4.
Gauge Pressure and Intravenous Fluids
A plastic bag for intravenous therapy contains a glucose solution. If the gauge pressure in a vein is 2000 Pa, what must be the minimum height of the bag to allow the glucose into the vein?

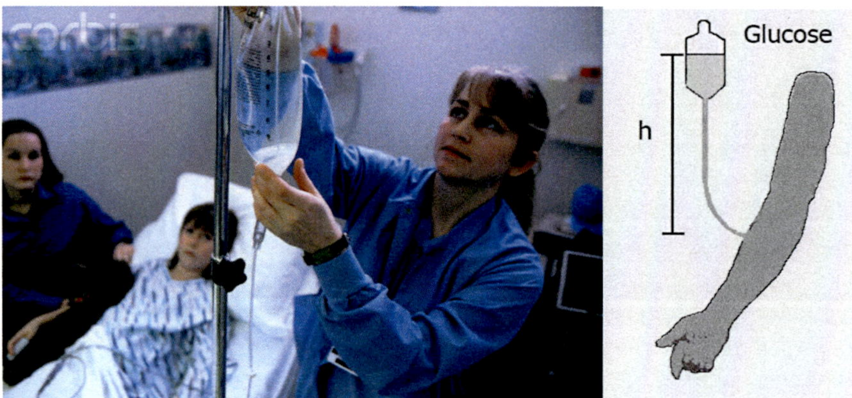

Solution

Pressure in the vein must be counterbalanced by the pressure due to the weight of the glucose above the vein. Or

Pressure in the vein = Pressure due to the weight of the glucose in the bag and tube above the vein = D × h × g.

Here D is the density of glucose (about 1000 kg/m^3), h is the height of the glucose solution in the bag and the tube above the vein, and g = 9.8 m/s^2.

2000 Pa = (1000 kg/m^3)(9.8 m/s^2)(h) or $$h = \frac{2000\ \text{Pa}}{9800\ \text{kg/m}^2\text{s}^2} = 0.2\ \text{m}$$

The nurse needs to place the glucose bag so that the glucose level is at least 0.2 m above the insertion point into the vein.

Force and pressure are closely linked. If you squeeze a water bottle with its top open, the water squirts out because the force of your hands created a pressure imbalance. The pressure in the bottle becomes higher than the pressure outside of it. A similar pressure imbalance forces water out of a showerhead to produce the hard spray. Water accelerates toward the lower pressure. This is the principle of a water pump that makes water move, often against the force of gravity.

An economical way to create water pressure is to use gravity. After all, gravity is everywhere, and it is free. The pressure of water in a vertical pipe increases by about 10,000 Pa for each meter of depth. Engineers in ancient Rome applied these principles when they built the advanced waterway system known as the aqueducts. We apply the same knowledge today when we design our community water towers. See **Figure 6.10**.

Gravity and water pressure Figure 6.10

a. The Roman civilization's highly advanced waterway system, known as the aqueducts, provided running water to the population. The entire system relied upon slopes and gravity to maintain water pressure and continuous water flow. The aqueducts had a gradual downward slope of 1/200 or less. When water had to cross large valleys, Roman builders constructed arcades, or arched bridges, to maintain the desired slope.

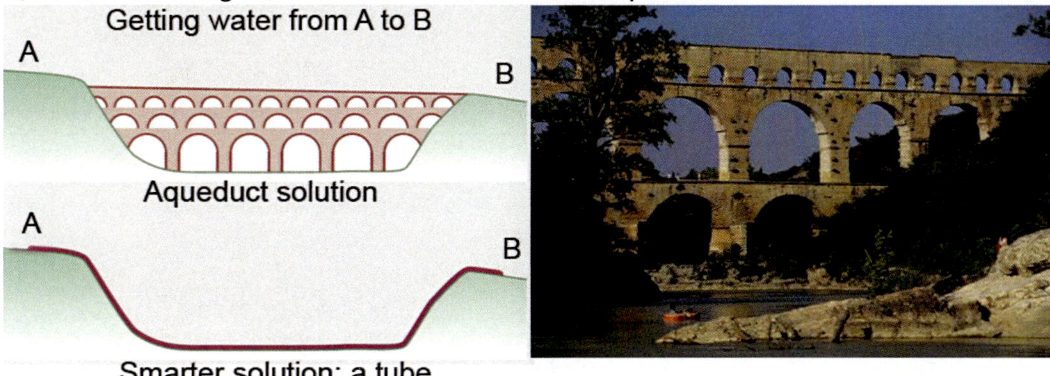

The Romans probably knew of the tube solution. The water will rise to the same level on the other side of the valley provided the pipe will stand the pressure at the bottom of the valley without leaking or exploding. The pressure at the bottom is around one atmosphere for every ten metres below the starting level the tube goes. Using the technology of the time it would be almost impossible to stop the pipe leaking if it was deep in the valley.

b. Many cities and communities have tall water towers, usually built on the highest site. As long as the tower is higher than the tallest building, the town has water pressure in all buildings.

A pump fills the water tower with water, and the water's weight provides a constant pressure throughout the plumbing connected to it. When the water level in the tower drops below a certain point, a pump refills the tower.

CONCEPT CHECK

1. **How** much does pressure change in an ocean for every 10 meters increase in depth?
2. **What** pressure do we measure when we check the blood pressure in our heart?
3. **How** does a medical syringe work?

6.3 Archimedes' Principle and Flotation

LEARNING OBJECTIVES

1. **Explain** buoyant force.
2. **Describe** Archimedes' principle.
3. **Explain** conditions for flotation.
4. **Define** surface tension.

If you have ever lifted a large rock or seashell from the bottom of a lake or the sea, you have noticed that the rock seems to weigh less while under water than it does when it breaks the surface. The water seems to counteract the downward force of gravity on the rock while it is submerged. In the same way, some objects can float in water in spite of the force of gravity that should be pushing them to the bottom. The ancient Greek scientist Archimedes studied this behavior, defining the upward buoyant force and determining the conditions for flotation. His insights led to Archimedes' principle, one of the most widely known physical laws today.

Archimedes' Principle

If you push down on a block of Styrofoam or wood floating in water, you can feel the water pushing back up on the block. A **buoyant** or upward **force** opposes the downward-pointing force. Similarly, you also feel lighter when walking in a swimming pool because the buoyant force opposes your weight. Many athletes keep fit by running in water, especially while recovering from knee or foot surgeries, because the stress on their feet in water is much lower than while running in air.

buoyant force An upward force due to the difference between forces acting on the top and bottom surfaces of an object immersed in a fluid.

To understand the origin of the buoyant force, imagine a solid cube of any material immersed in water. The pressure on the top face is smaller than the pressure on the bottom face because pressure increases with depth. See **Figure 6.11**. The buoyant force on the cube is the resultant of the two opposite forces acting on the top and bottom faces of the cube. The resulting force points upward. Buoyancy acts on objects of any shape, not just cubes.

How can we determine the magnitude of the buoyant force on a given object? Let's start with a simple experiment. If you drop an object such as a stone in any container of water, the water level rises because the stone pushes out and replaces the water. The amount of water displaced equals the volume of the stone.

The connection between the buoyant force acting on an object and the amount of water displaced by the object is called **Archimedes' principle**. To better understand buoyant forces and derive **Archimedes' principle**, let's consider using a spring scale to measure the weight of an object in air and the weight of the same object while submerged in water (apparent weight). The two measurements link the buoyant force to the weight of the water displaced. See **Figure 6.12**. **Example 6.5** applies this concept to an interesting challenge question.

Buoyant force acting on a solid object submerged in water Figure 6.11

The upward buoyant force on the cube is the difference between the two forces acting on the top and bottom faces of the cube. (Forces act on the sides of the cube, too, but they are equal and opposite, so they cancel out.)

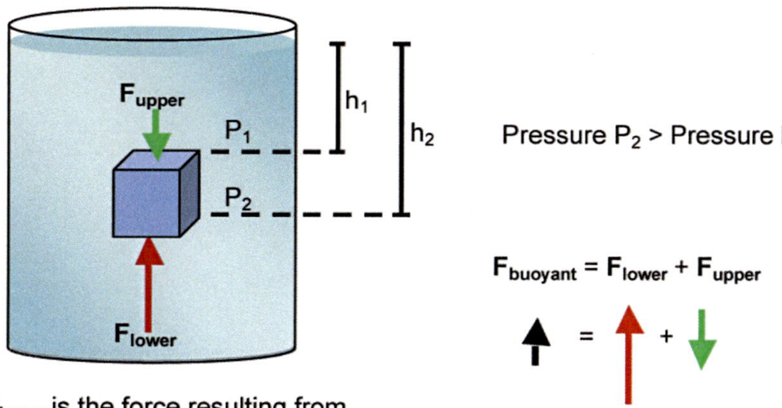

Archimedes' principle An object partially or wholly immersed in a fluid is acted upon by an upward buoyant force equal to the weight of the fluid displaced by the object.

Buoyant force and Archimedes' principle Figure 6.12

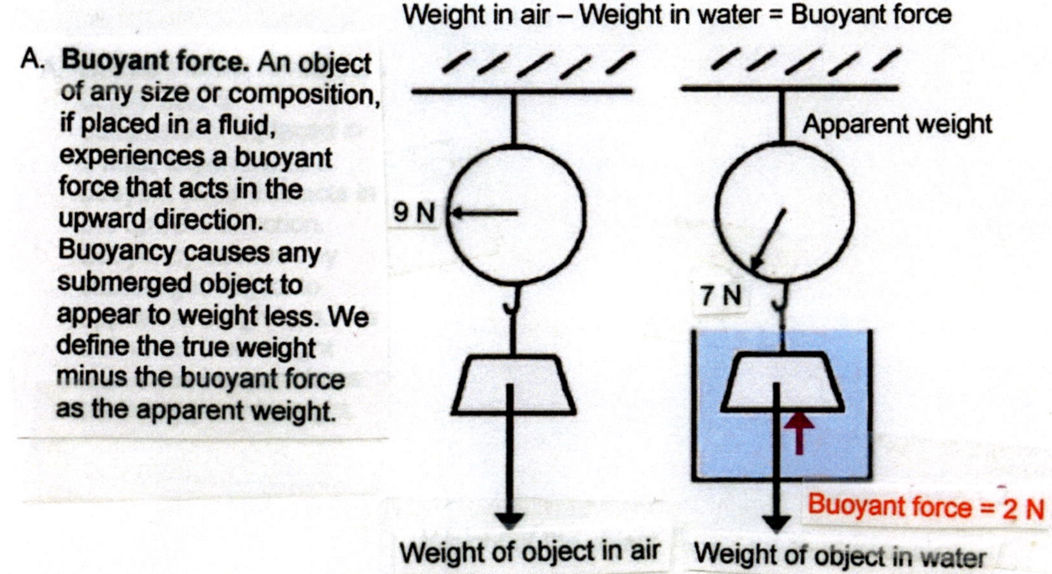

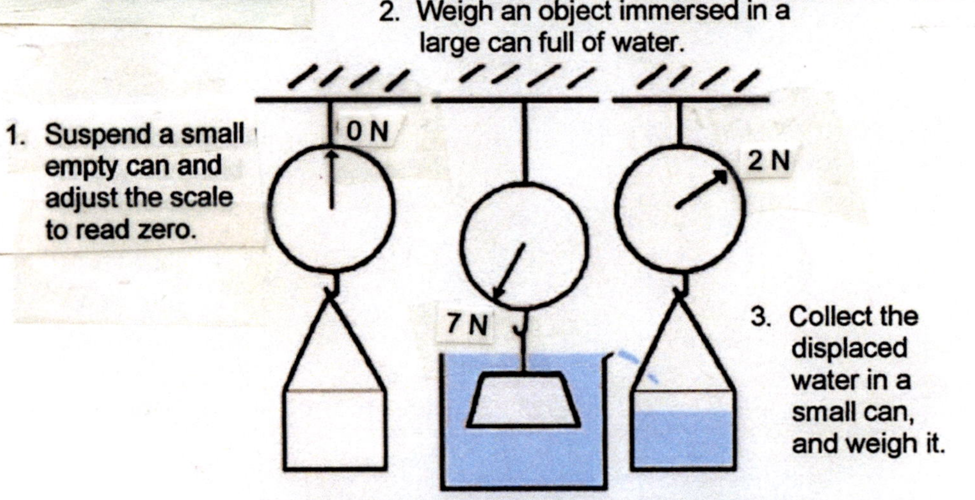

Example 6.5
Water Displacement
An inquisitive student is in a boat in a swimming pool. Also in the boat is a rock. The student picks up the rock and tosses it into the pool, where it sinks to the bottom. No water leaves the pool from the splash made by the rock. The student asks a question: Does the pool's water level rise, lower, or stay the same?

Solution:

Let's examine all three possibilities:

If you guessed that the pool's water level rises, it's not the right guess. You may have thought that the rock displaces more water when it's in the pool than when it's in the boat. But does it?

If you guessed that the pool's water level stays the same, this guess is also not right, although this is the answer that most people choose. You may have thought that the rock displaces the same amount of water if it's in the boat or in the water. But it doesn't!

Finally, if you think that the pool's water level becomes lower, you are right! The reason is that when the rock is in the boat, it displaces its total weight. The extra buoyant force needed for the boat to float while carrying the rock equals the weight of the rock. A rock weighing 50 newtons in the boat must displace a volume of water that weighs 50 newtons.

When the rock is sitting at the bottom of the pool, it displaces its own volume. A given volume of rock is typically 3 to 5 times heavier than the same volume of water. The 50-newton rock may displace a volume of water that weighs only 15 newtons.

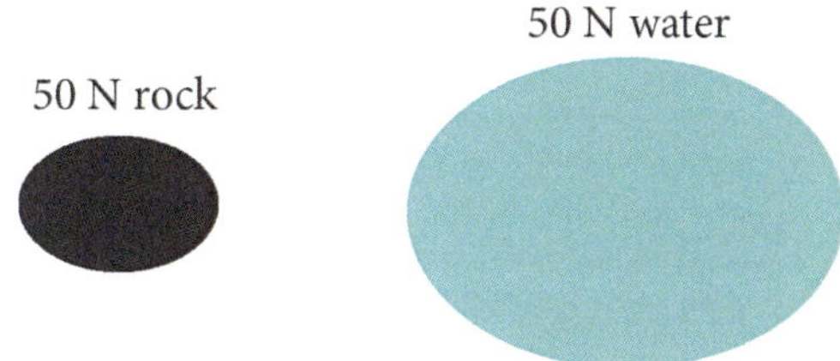

The rock displaces more water when it is in the boat than when it is in the water, so the pool's water level is lower when the rock is in the water.

Flotation

You can apply Archimedes' principle to find out if an object can float. Flotation is an important concept for those who design steel tankers, life jackets, or even America's Cup racing yachts. These designers need to know under what conditions an object can float and how the object's overall density affects flotation.

Consider a huge Styrofoam slab. It is massive, but it floats. Styrofoam's density is less than that of water. Therefore, you may conclude that if an object is less dense than a fluid, it will float in that fluid. The resultant force acting on a floating object must be zero; otherwise it would rise or sink. Thus, *while floating, an object's weight is equal to the buoyant force,* which equals the weight of the water displaced by the object. See **Figure 6.13**. If the object is denser than the liquid, it will sink because its weight is greater than the buoyant force. **Example 6.6** gives a numerical example.

Flotation condition and buoyant force Figure 6.13

a. Floating ice cube is supported by buoyant force that equals the weight of the ice cube.

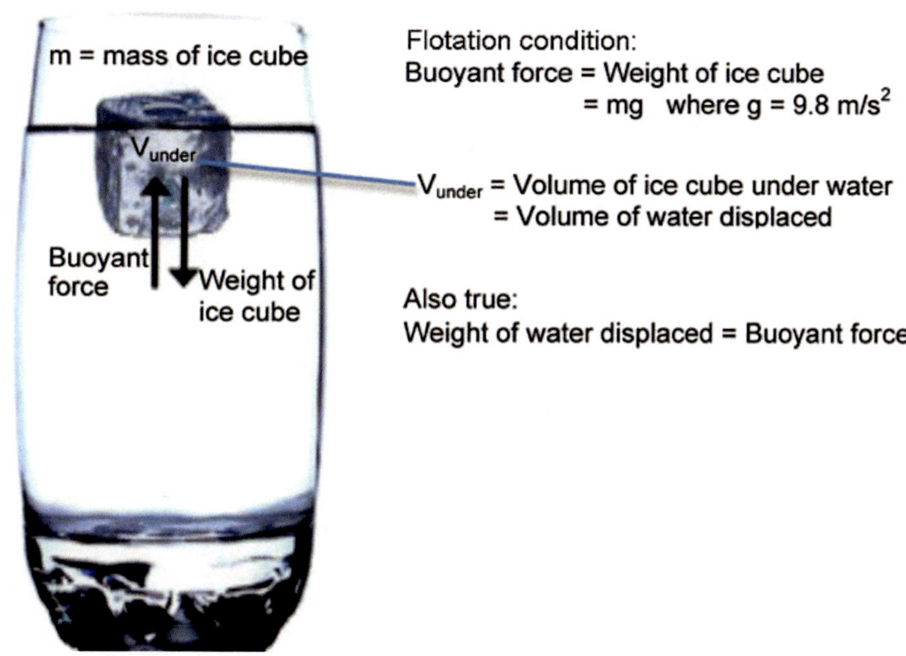

b. The weight of the water displaced by a floating object is equal to the weight of the object.

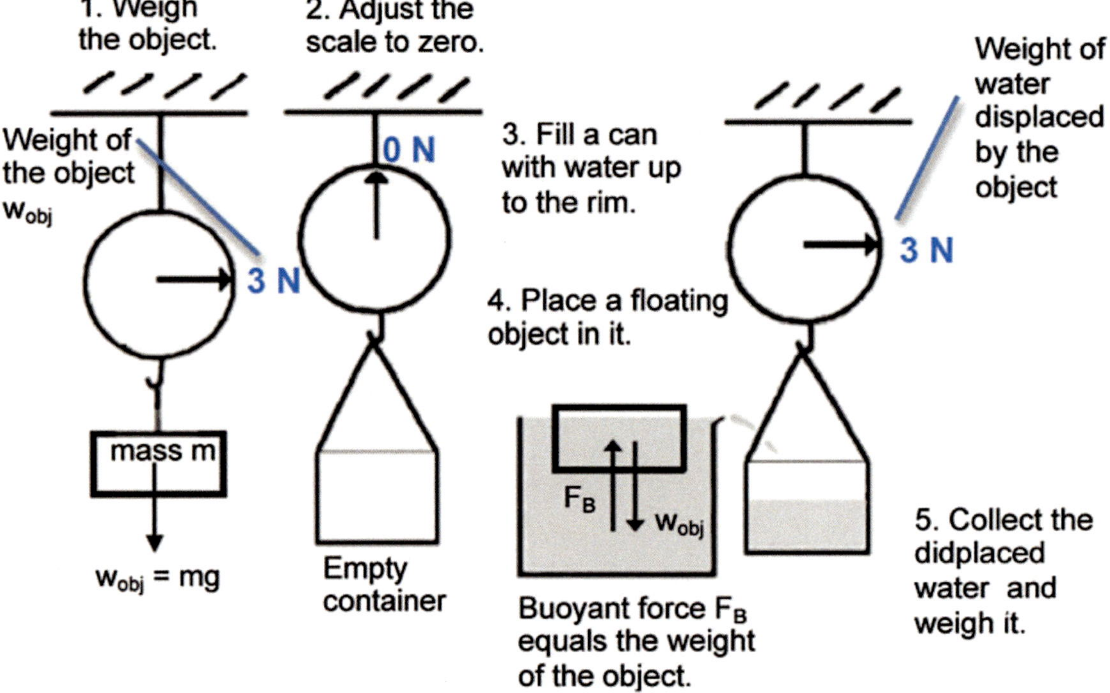

Example 6.6
Buoyancy of Life Jackets
A survivor from a sinking ferry floats on the sea near Karimun Island, Indonesia, before being rescued on November 22, 2009.

Life jackets are rated by their buoyancy. What is the buoyant force on a floating life jacket if 0.01 m^3 of the jacket is normally submerged under water?

Solution:
The total buoyant force is the sum of the buoyant forces due to the water displaced by the person and the water displaced by the life jacket. The total weight of the person and the life jacket must be balanced by the total buoyant force to have flotation.

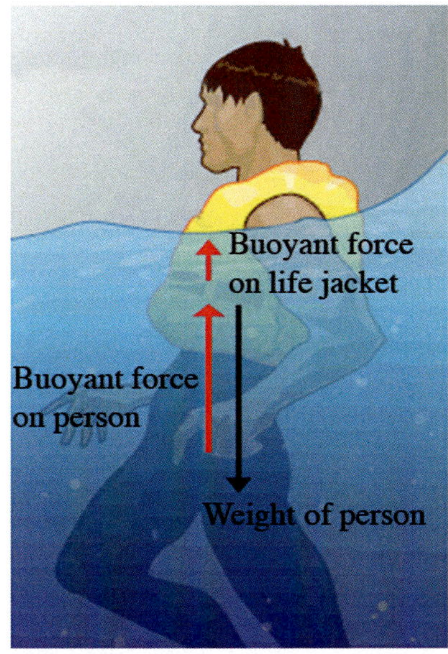

Without a life jacket a still person in water would sink because his weight would be larger than the buoyant force. Here we neglect the weight of the life jacket.

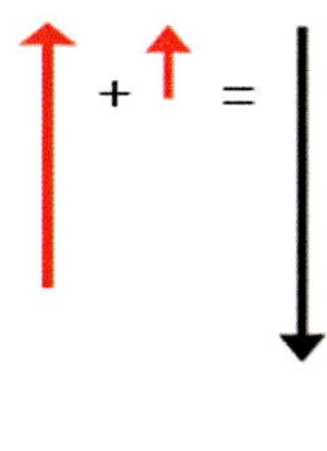

If 0.01 m^3 of the jacket is under water, then it displaces 0.01 m^3 of water. We know that water's density is 1000 kilograms per cubic meter. Thus the jacket displaces

$$1000 \text{ kg/m}^3 \times 0.01 \text{ m}^3 = 10 \text{ kg of water.}$$

The weight of the displaced water (mass times the acceleration due to gravity) is

$$m \times g = 10 \text{ kg} \times 9.8 \text{ m/s}^2 = 98 \text{ N}$$

Therefore, the buoyant force on the life jacket, which equals the weight of the displaced water, is 98 N.

The U.S. Coast Guard requires that a type I personal flotation device have a buoyancy of at least 98 N. Average adults require only 35-55 N of buoyancy to keep their heads above calm water. Thus 98 N of buoyancy should be plenty to keep your head out of water.

Buoyancy and flotation have many interesting applications. The sinking of the *Titanic* is an example of water displacement that shows how the ship's increasing weight and decreasing buoyant force sank the ship. See **Example 6.7**.

Example 6.7
Sinking of the Titanic

The Titanic sank on April 15, 1912, after it hit a large iceberg. The ship's hull was designed to displace 66,000 metric tons of water. (One metric ton is equivalent to a mass of 1000 kg and a weight of 9,800 N.) The ship's gross mass was 46,328 tons, or 46,328,000 kg. Therefore, the Titanic had 19,672 tons (66,000 tons - 46,328 tons) of extra capacity. This is equivalent to extra weight capacity (displacement capacity) of 19,672,000 kg $\times$ 9.8 m/s^2 = 192,785,600 N. If the hull was divided into 16 compartments separated by 15 watertight bulkheads, how many compartments needed to be filled for the Titanic to sink? In this example, we use ton units because that's what the shipbuilding industry uses.

Solution:
If the ship lost more than 19,672 tons of its displacement capacity, the gross weight would exceed the buoyant force, and she would sink. Assuming the hull was divided into 16 compartments separated by 15 watertight bulkheads then each compartment of the hull had the same displacement capacity: 4125 tons (66,000 tons/16 compartments).

The Titanic sank once the collision with the iceberg ripped open at least five compartments because 19,672 tons – 5 × 4125 tons = 19,672 – 20,625 = –953 tons.

At that point the weight of the ship was greater than the buoyant force supporting it. The Titanic's bow flooded first, so it sank nose-first. You can see an animation of this sinking on www.youtube.com. Use the search words "Titanic sinking."

Other examples of flotation include the way fish and submarines move. Fish move up and down within the water by manipulating how much air they hold to change their densities. See **Figure 6.14**. Submarines use both water and air to change their density. Submarines cruising at a definite depth in the ocean have an average density equal to the density of ocean water, 1025 kg/m^3. This means that the density of a submarine cruising at a given depth—total mass (metal shell, air, crew, load, etc.) divided by the total volume—is 1025 kg/m^3. To descend, the crew allows water into the ballast tanks, so the submarine's density increases. To rise, they force compressed air into the tanks to expel water and decrease the submarine's density.

You have studied the basic principle of flotation, but saying that something floats does not give a complete picture. If you ever visited a harbor, did you notice that water came up to different points on similar boats? Similar-looking objects may have different water lines. Examine a floating ice cube. Most of it sits below the water's surface. However, more of a Styrofoam cube with an identical shape will sit above the water line. What determines how much of an object is submerged under water?

Increasing the density of a floating block by adding mass to it causes the block to sink lower into the water. Eventually the density increases enough for the block to submerge completely. At this point the block displaces the maximum amount of water and experiences the maximum buoyant force possible; however, the buoyant force cannot counterbalance a larger weight. When the buoyant force and the weight of the block are equal, the block becomes, in a sense, weightless in the water. Adding a small mass to the block to make it slightly denser causes the block to sink to the bottom. Interestingly, more muscular swimmers are less likely to float than swimmers with more body fat because muscle is denser then fat.

Regulating depth with air and water Figure 6.14

Fish are denser than water. Then how do they regulate their depth? Fish have air bladders, which are thin-walled sacs. The fish's gills collect air dissolved in the water. The fish regulates the size of its bladder to move up toward the water's surface or sink lower in the water.

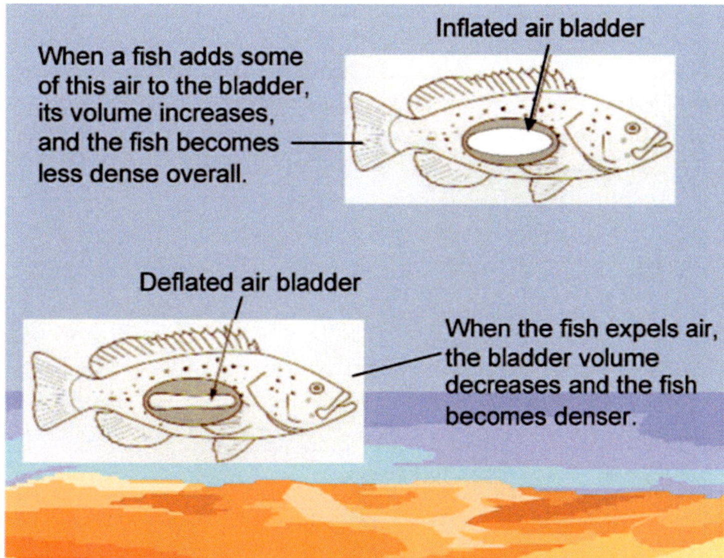

Another factor that affects flotation is the density of the liquid in which the object floats. If you travel to the Dead Sea, which is known for its high salt content, you will be amused by being able to lie still on its surface without sinking. See **Figure 6.15**.

Floating in the Dead Sea Figure 6.15

You can experience an extreme flotation effect in the Dead Sea, which is an inland body of water southeast of Jerusalem on the border of Jordan and Israel. It has the lowest elevation on Earth—more than 400 m below sea level at the water's surface. Because it loses water only by evaporation, the Dead Sea has a high salt content: more than 25 percent by weight, or seven times that of ocean water. The water in the Dead Sea has an average density of 1.24 kg/L, much higher than normal sea water. Its buoyancy makes it impossible for a swimmer to sink. You can lie down in the water and relax, without needing to move to keep your body above the surface.

Surface Tension

Have you ever observed insects, such as water striders, standing on the surface of a pond? They are not partially submerged, and therefore no buoyant force acts on them. A leg of an insect can make a small circular depression in the surface of the water. See **Figure 6.16a**. It is as if an elastic sheet covers the water's surface. We say that the water's surface has **surface tension**.

surface tension Intermolecular forces in liquids that create a surface membrane.

A useful analogy for surface tension is to think of a tightrope walker supported by a horizontal rope with some tension in it (**Figure 6.16b)**. The vertical weight of the person is opposed by the vertical components of the tension. In two dimensions, the strong, attractive intermolecular forces in water create a "skin" on the liquid's surface (**Figure 6.16c**).

Surface tension Figure 6.16

a. For tiny objects like this water strider, the water's surface acts like an elastic membrane.

b. A slackliner is supported by a horizontal rope with some tension in it. The rope bends where the performer stands.

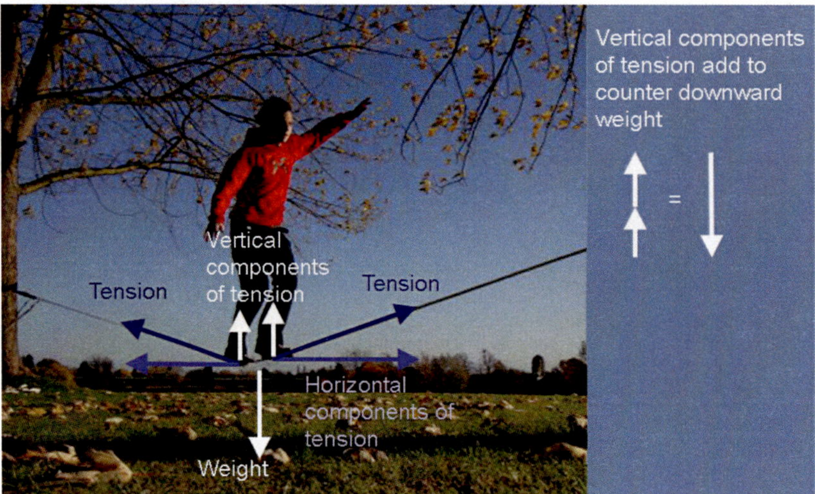

c. Surface tension on the molecular level.

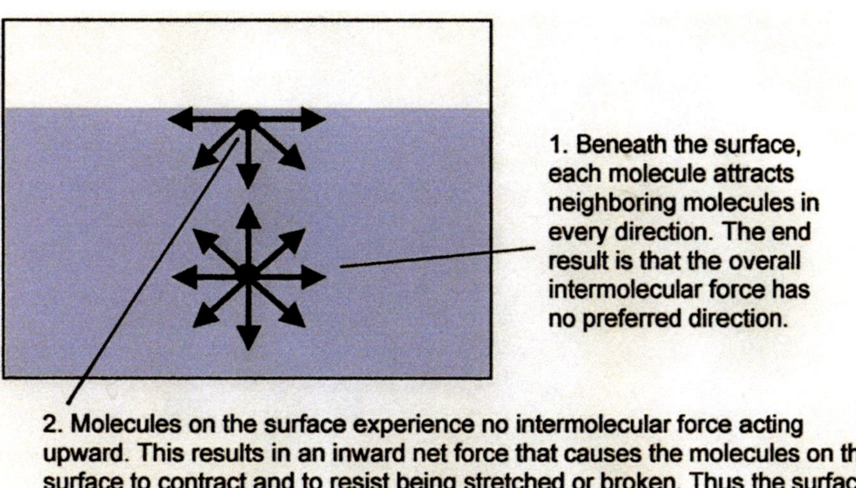

The tendency for a liquid to pull the surface molecules inward causes the surface area to be as small as possible, because it takes energy to stretch this elastic membrane. This is why small raindrops (less than 1 mm) have spherical shapes. Surface tension seeks to arrange the shape of a raindrop such that it is in the lowest energy state. Therefore, raindrops seek the smallest surface area for a given volume, which is a sphere. In practice, raindrops take on other shapes because the air pressure pushing up against the bottom of large drops tends to flatten them as they fall.

The surface tension of water is larger than that of any other common liquid except mercury. This is the reason why we use soapy hot water to wash our clothes better. Soapy hot water helps clean clothes by lowering the surface tension of the water so that it more readily soaks into pores and soiled areas. See **Figure 6.17**.

Lowering surface tension by adding detergent Figure 6.17
The surface of the soapy water is attracted more strongly to the surface being cleaned than to the water itself. Consequently, the water spreads out along a solid surface rather than beading upon it. The soap is said to make water wetter.

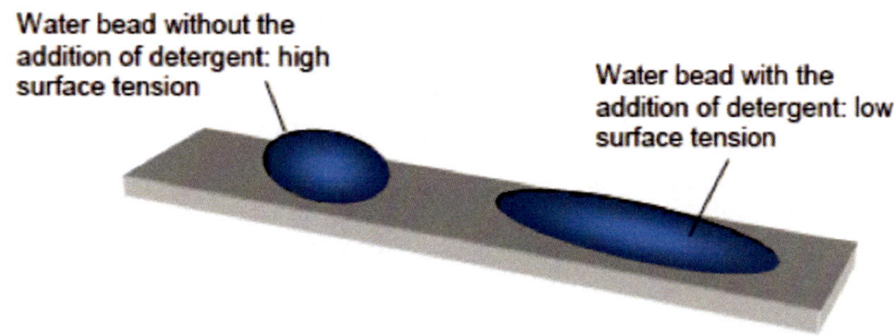

The high surface tension of water is responsible for the formation of water beads on this stem of grass.

CONCEPT CHECK

1. **What** is the cause of buoyant force?
2. **How** does weight of the displaced water compare to the buoyant force on an object immersed in water?
3. **What** are the conditions under which metals can float?
4. **Why** does soap lower surface tension in water?

6.4 Earth's Atmosphere

LEARNING OBJECTIVES

1. **Describe** properties of the atmosphere.
2. **Explain** the atmospheric pressure.

So far we learned about liquids and solids. Here we will study the gaseous state, the third and the final state of matter on Earth. The most common gas, familiar to everyone, is air, consisting mostly oxygen and nitrogen. Oxygen in air is essential for life of all humans, animals, and plants. Without it we would not have fire. Nitrogen, on the other hand, does not support life or fire.

Life as we know it would be impossible on Earth without the oxygen that we breath. Oxygen is a part of the layer of gases held close to the surface of a planet. We call this gas layer our *atmosphere*. In addition to supplying the oxygen to sustain life, the atmosphere also shields us from harmful radiation coming from the Sun, generates rain-giving clouds and traps just enough heat to keep our planet livable.

Atmospheric properties

The atmosphere of the Earth is a mixture of gases. Dry air is composed of approximately 21% oxygen, 78% nitrogen, and 1% argon by volume in addition to tiny traces of other gases. However, water on the surface of the Earth evaporates so there is a small amount of water vapor present too. Water vapor accounts for about one-quarter of 1 percent of the total volume of the atmosphere. The gas molecules in atmosphere have kinetic energy and move in a Brownian like random motion. Without the gravitational pull of the Earth, the molecules would have no reason to stay around the Earth and would spread on into outer space.

There is no distinct boundary where our atmosphere stops. The concentration of gases is greatest at the sea level and decreases steadily with altitude. Eventually it empties out into space. The density of the air decreases as we move up in altitude. At sea level, 1 m^3 has mass of about 1.25 kg. At an altitude of 3 km (10,000 feet), the density drops by 26% compared to sea level. At 6 km (20,000 feet) the density drops by almost a half the value at sea level. The density of air at the top of Mt. Everest is about only 30% of the normal air density that we are accustomed to.

More than 99% of the total mass of the atmosphere is found below 40 km (25 miles) above sea level. Relative to the size of the Earth, it is a thin layer. See **Figure 6.18**. If the Earth were a ball with a radius of one meter, breathable air would extend only one millimeter outward. Planets such as Mars with a small mass and a weaker gravitational attraction have even thinner atmospheres than ours, and small bodies such as our moon cannot retain a gaseous shell at all.

Density of atmosphere at different altitudes Figure 6.18

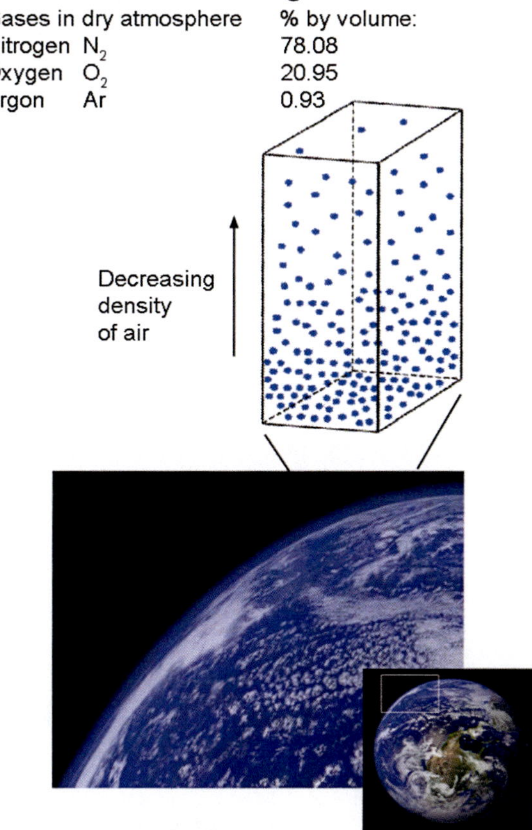

Atmosphere forms a thin layer of gases around the Earth. More than 99% of the total mass of the air is below 40 km (25 miles). The density of air steadily decreases with altitude. At 40 km it is less than 1% of its value at sea level.

Atmospheric pressure

We live under the weight of an ocean of air. This air exerts pressure on us just like water exerts pressure on a diver. We call it the **atmospheric pressure**. We can model this pressure by imagining a tall column of air, having a square base of 1 m^2 and reaching 40 km high. The weight of the air inside the column is about 101,000 N and this weight produces pressure of 101,000 N/m^2 at sea level. See **Figure 6.19**.

Atmospheric pressure Figure 6.19
1 m^2 column of air, 40 km high, has mass of 10,000 kg. Its weight causes atmospheric pressure.

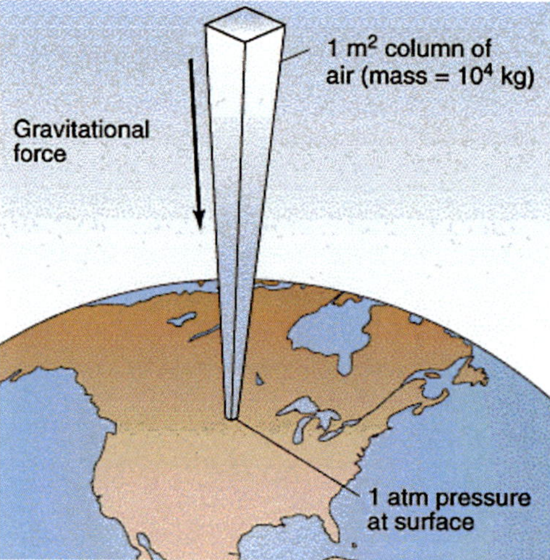

atmospheric pressure The pressure exerted on us by air in the atmosphere. At sea level this pressure is about 101,000 Pa.

The atmospheric pressure at sea level produces large forces, including the force acting on you. If the surface area of your body is about 2 m^2, then your body experiences a force of about 202,000 N! This is a weight equivalent to mass of more than 20,000 kg or about 44,000 lbs weight! Your body reacts with an equal and opposite force so it balances this force from outside and therefore does not feel it.

During the 1600s, a German scientist, Otto von Guericke, showed the huge force of the atmosphere in a dramatic experiment. He placed together two copper bowls 36 cm (14 inches) in diameter, with a good seal in between to form a hollow sphere. After the air was pumped out, even a team of 16 horses, 8 on each side, pulling on each hemisphere, could not pull them apart. See **Figure 6.20**. Had he understood Newton's 3rd law, he could had created the same force by tying one end of the rope to a sturdy house and the other end to 8 horses.

Magdeburg hemispheres Figure 6.20
After pumping the air out of the hemispheres, the pressure inside drops. The inward pointing force due to the external atmospheric pressure dominates and it becomes very hard to separate the hemispheres.

Initially

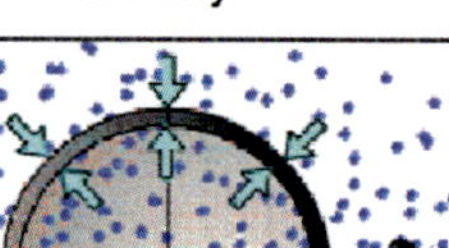

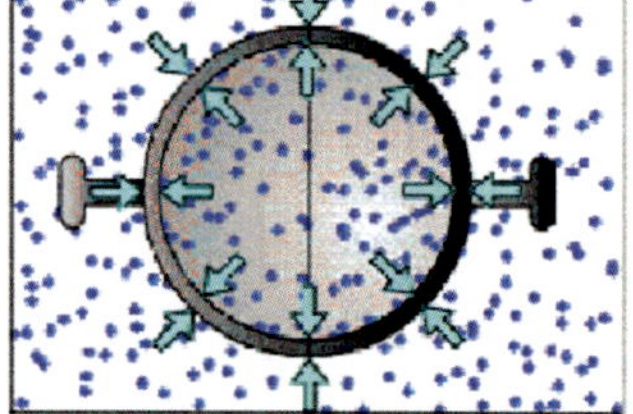

The force due to air pressure is the same both inside and outside the hemispheres.

After pumping air out

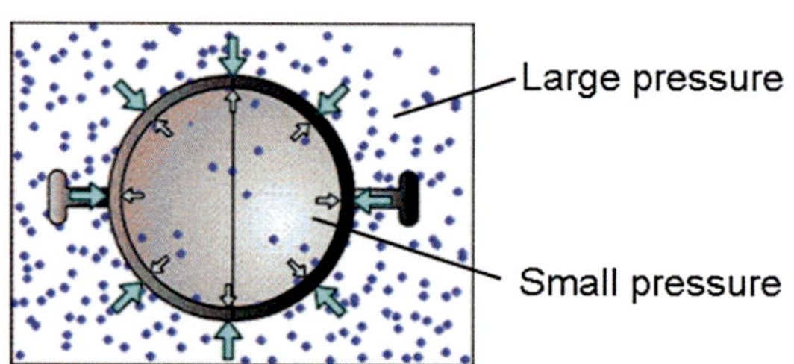

The outward force inside the hemispheres is smaller than the inward force due to atmospheric pressure.

During tornados or hurricanes, the problem is not an increased air pressure but rather a drop in pressure as we will learn later in the chapter. **Example 6.7** shows what a small drop in the atmospheric pressure can do to a garage door.

Example 6.7
Pressure on a garage door
A small, yet significant, pressure drop could occur in a tornado. Calculate the net force on a garage door of a closed house if the outside pressure suddenly dropped 10 percent. The dimensions of the door are 2.40 m (8 feet) times 3.60 m (12 feet).

The atmospheric pressure is equal to 101,000 N/m^2. A 10% drop in the atmospheric pressure is approximately equal to 10,000 N/m^2. Surface area of the door = (2.40 m)(3.60 m) = 8.64 m^2. Change in force on the door = (change in pressure) $\times$ (surface area of the door) = (10,000 N/m^2)(8.64 m^2) = 86,400 N. This force will point outward and is equivalent to the weight of an 8,816 kg mass! A pressure drop much smaller than this can cause significant damage to our homes.

Mountain climbers trying to reach Mt. Everest face serious reduction of pressure too. The air pressure at the top is only about 31,000 Pa. Most of us are not able to function at air pressures so low and many climbers have lost consciousness and died in the extreme cold of Mt. Everest. See **Figure 6.21**. Large airplanes fly at altitudes of about 10-12 km. Therefore, airplanes must be furnished with special atmospheric control systems that pressurize cabins. For instance, at a cruising altitude of 12,000 meters (40,000 ft), a Boeing 767's cabin will be pressurized to an altitude of 2,100 meters (7,000 ft), a level acceptable to our bodies.

Air pressure on Mt. Everest Figure 6.21
As one ascends through the atmosphere, the air pressure decreases, and every breath contains fewer and fewer molecules of oxygen. One must work harder to obtain oxygen, primarily by breathing faster. The blood becomes less and less efficient at acquiring and transporting oxygen. This means that no matter how fast one breathes, attaining normal blood levels of oxygen is not possible at high altitude.

CONCEPT CHECK

1. **How** high can we climb before we develop breathing problems and faint due lack of oxygen?
2. **Why** are two connected suction cups hard to separate?

6.5 Ideal gas and its properties

LEARNING OBJECTIVES

1. **Explain** temperature.
2. **Describe** ideal gas.
3. **Define** ideal gas law.

One liter of air contains about 10^{22} molecules, a very large number. Therefore, it is not possible to describe gas one molecule at a time and we use macroscopic quantities that take into account gas properties on a larger scale – pressure, volume, temperature, and the amount of gas. We will describe temperature, the ideal gas law, and study how gases respond to changes in pressure, volume, and temperature, while keeping the amount of gas constant. Hot air balloons or scuba diving will provide some interesting applications.

Gas has neither a definite shape nor a definite volume. Often it is not visible to us. Clouds, made out water vapor, are one exception. We will examine the four important properties that describe gas – pressure, volume, temperature, and density of gas, and see how they are linked by the ideal gas law. They explain many phenomena that we see and use in our lives.

Temperature

Temperature is one of the more familiar physical concepts in our lives. It tells us how hot or cold an object is. But what is really behind temperature? All matter is composed of continuously moving or just jiggling atoms or molecules which possess kinetic energy.

A gas molecule undergoes up to 2 billion collisions each second. Each collision changes the molecule's speed. As a result, the gas has a broad distribution of speeds. In addition, the average speed increases with temperature (**Figure 6.22**). At room temperature, the average speed is about 500 m/s. But because of the constant collisions, an individual gas molecule does not get too far. Its net displacement is a small fraction of 1 mm each second.

Distribution of molecular speeds at different temperatures Figure 6.22
Molecular speeds spread over broad range. At the same time the average speed increases with temperature.

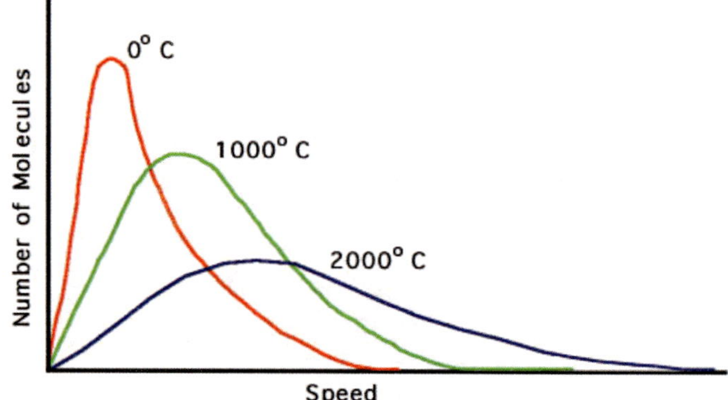

The speeds here range from 0 to 4 km/s.

temperature A measure of the average kinetic energy per molecule in some substance

The sum of the average kinetic energies of all molecules is a measure of what we would call warmth. In general, the greater the temperature of a substance, the greater is the motion of its molecules. A relationship can be derived that links temperature with the average kinetic energy of the gas molecules. *Temperature is a measure of the average kinetic energy, KE_{ave}, of the molecules of a substance.* In symbols,

$$KE_{ave} = (\frac{1}{2}mv^2)_{average} = \frac{3}{2}kT .$$

Here m is a mass of a molecule, v is its velocity, k is Boltzmann's constant = 1.38 10^{-23} m^2 kg s^{-2}, and T is the absolute temperature measured in kelvins. This relation states that the average kinetic energy of substance (per molecule) is proportional to the temperature of the substance. This means that more massive molecules such as oxygen will have lower average velocity than lighter molecules such as hydrogen at a given temperature.

We have not formally defined temperature. How do we specify and measure temperature? Our perception of touch is not adequate to determine temperature well. Yet, no efforts to quantify temperature were documented until the late 16th century. Galileo invented a crude thermometer in 1592. Today all thermometers depend on some physical property that changes with temperature. The most common type, including the one first proposed by Galileo, is based on liquids property to expand or contract when heated or cooled. The liquid, usually mercury or red-colored alcohol, rises or falls in a glass tube as temperature changes.

The most widely used temperature scale assigns the number 0 to temperature at which water freezes and number 100 to temperature at which water boils (at standard atmospheric pressure). The space between the two points is divided into 100 equal sized degrees. Each degree is called *Celsius (°C)* degree in honor of Andres Celsius, a Swedish astronomer, who first proposed this scale in 1742.

In United States, we use the *Fahrenheit* scale. On this scale we assign the number 32 to temperature at which water freezes and 212 to temperature at which water boils. There are 180 divisions, or Fahrenheit degrees, between freezing and boiling points of water. Therefore, the size of a Celsius degree is 9/5 larger than the size of a Fahrenheit degree. To convert from Fahrenheit to Celsius and vice versa we use the following conversion formulas:

$$T_F = 32 + (9/5)\ T_C \quad \text{and} \quad T_C = 5/9\ (T_F\text{-}32)$$

Both temperature scales have temperatures that can be positive or negative numbers. Scientist thought it would be easier to find the lowest possible temperature and assign it value of 0. This would give us positive numbers for all temperatures. But what is this lowest temperature? For a long time nobody had the answer.

In 1848, William Thomson (also known as Lord Kelvin) proposed the **absolute temperature scale**. He measured volume of gases as their temperature was lowered at constant pressure. The data fitted a straight line. There was a limitation to this measurement. Eventually the gas would condense. This would happen at about –200°C. Since the volume-temperature graph fits a straight line, one can extrapolate the line until it intercepts the temperature axis. This temperature intercept is –273.15°C. This point is the very limit of low temperature and is defined as the absolute zero or 0 K. We obtain temperature in Kelvin degrees by adding 273.15 to temperature in Celsius degrees. Each Kelvin degree has the same size as Celsius degree.

absolute temperature scale Temperature scale that uses Kelvin degrees and has 0 at –273.15°C. One converts to Kelvin degrees by adding 273.15 to Celsius degrees.

Normal temperatures of the world around us vary from 4 K for liquefied helium to 6000 K at the surface of the Sun. The temperatures within the Sun are significantly higher, about 10,000,000 K.

Ideal gas properties

Gas is called **ideal** if its molecules are so far apart that they exert forces on each other only when they collide. The molecules travel freely in a straight-line motion until they hit each other or hit the walls of their container. The wall turns the molecules back into the body of the gas. The collisions are elastic. The directions of the particles velocities are random. The forces of the collision between the molecules and the wall form a Newtonian action-reaction pair. The molecules act with a force on the wall and the wall acts with an equal but opposite force back on the molecules. If this outward force on the wall is divided by the area of the wall, we obtain the pressure of the gas. See **Figure 6.23**.

Motion of gas molecules Figure 6.23

a. A gas may be pictured as a collection of widely spaced molecules in a continuous, random motion.

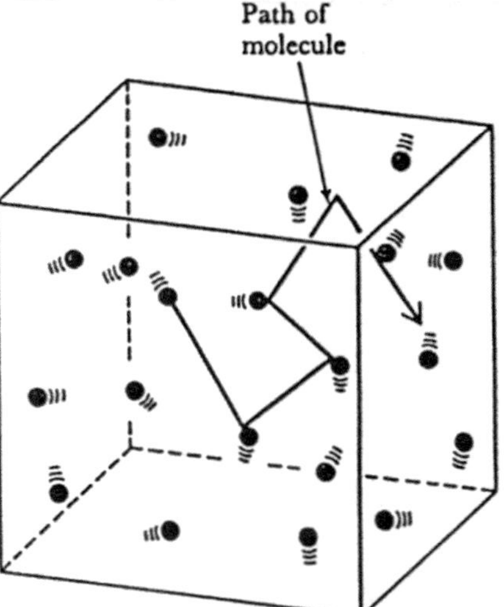

b. Just as a ball exerts force when it bounces against a wall, so do gas molecules exert force when they collide with a surface. These collisions sum up to cause pressure and constant force acting on the surfaces exposed to the gas.

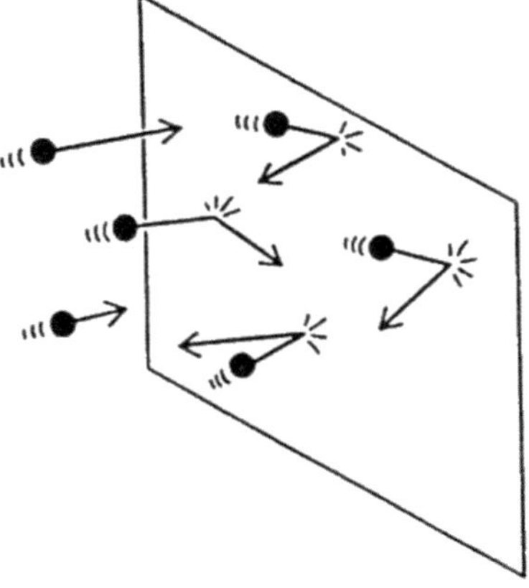

If we increase the amount of gas, and therefore its density, in some space, the number of collisions with the walls will increase. This increase will result in the increase of pressure acting on the walls. See **Figure 6.24**.

Pressure's dependence on density Figure 6.24
The total pressure exerted by the gas depends on the concentration of the gas molecules in the enclosed space. The higher the density, the higher is the pressure.

a.

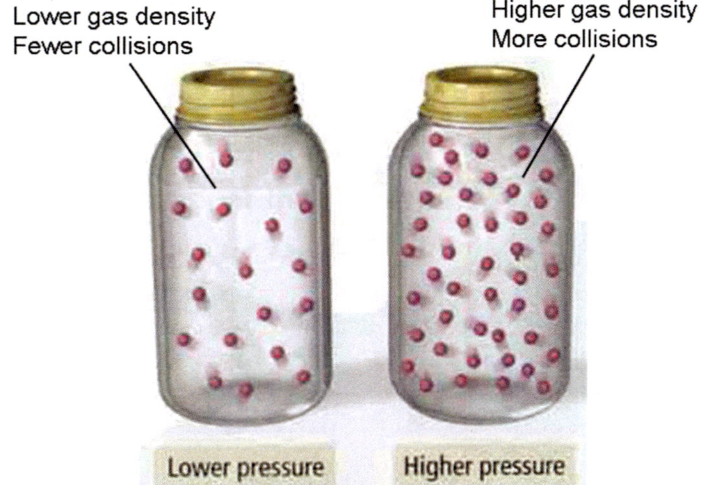

Anytime you inflate a bicycle tire, you are increasing its pressure by pumping more air inside.

b.

ideal gas Amount of gas that consists of a large number of molecules separated by large distances relative to their sizes, randomly colliding with each other, and without any inter-molecular forces present.

Ideal gas law and applications

We have learned that we can increase pressure by increasing density of gas. Another way to increase pressure is to compress the gas by decreasing volume. If you compress a gas, its molecules move closer together, and there is less open space between them. This increases the number of collisions of the gas molecules with any surface in their way, and the overall gas pressure increases. *Reducing the volume of a gas increases its pressure if the temperature and the amount of gas are constant.*

If we halve the volume, the pressure will double. If we reduce the volume to one-third, the pressure will triple and so on. The amount of the pressure increase is directly proportional to the amount of volume decrease. This means that at constant temperature, the product of volume and pressure for a given amount of gas is constant:

$$\text{Pressure} \times \text{Volume} = \text{constant}$$

Or in symbols,

$$PV = \text{constant (for a given amount of gas held at constant temperature)}$$

This relationship is known as *Boyle's law*. See **Figure 6.25a**.

Process Diagram Figure 6.25
Experiments leading to the ideal gas law

a. Boyle's law

We insert different weights onto a movable piston to create pressure. Pressure on the gas is proportional to the weight placed on the piston.

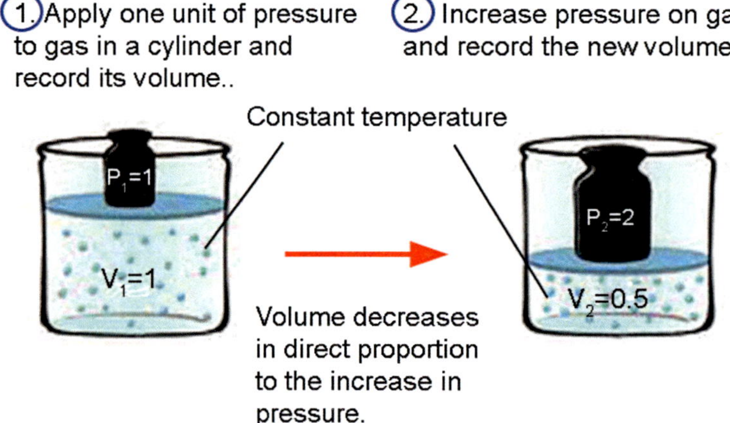

The product of volume and pressure for a given amount of gas is constant:

$$P_1V_1 = P_2V_2$$

Rising air bubble in water is a demonstration of Boyle's law. As the bubble rises, its volume increases while the pressure acting on it decreases.

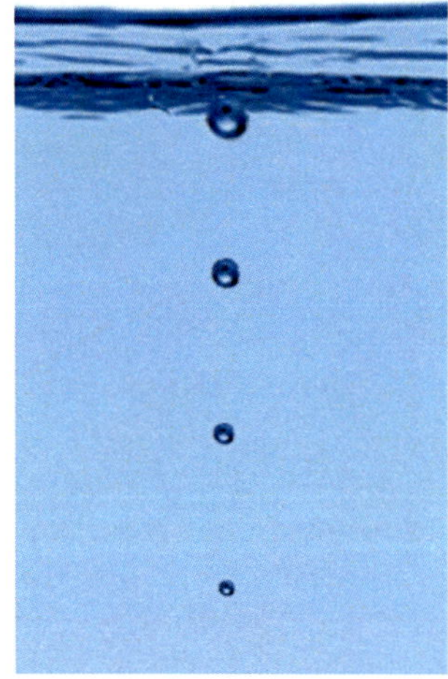

What is the effect of changing temperature on volume of gas kept at constant pressure? Place an inflated balloon in a freezer for a few minutes. What happened? It shrunk! Bring it back to room temperature. What happened now? It returned back to its original shape after a few minutes. This experiment shows that volume of gas changes with temperature. Jacques Charles (1746-1823), a Frenchman, first performed this experiment. He showed that volume of gas V is directly proportional to its temperature T or equivalently, the ratio of volume to temperature is constant at constant pressure. We call this relationship *Charles's law* (**Figure 6.25b**). In symbols,

$$\frac{V}{T} = \text{constant (for a given amount of gas held at constant pressure)}$$

b. Charles's law

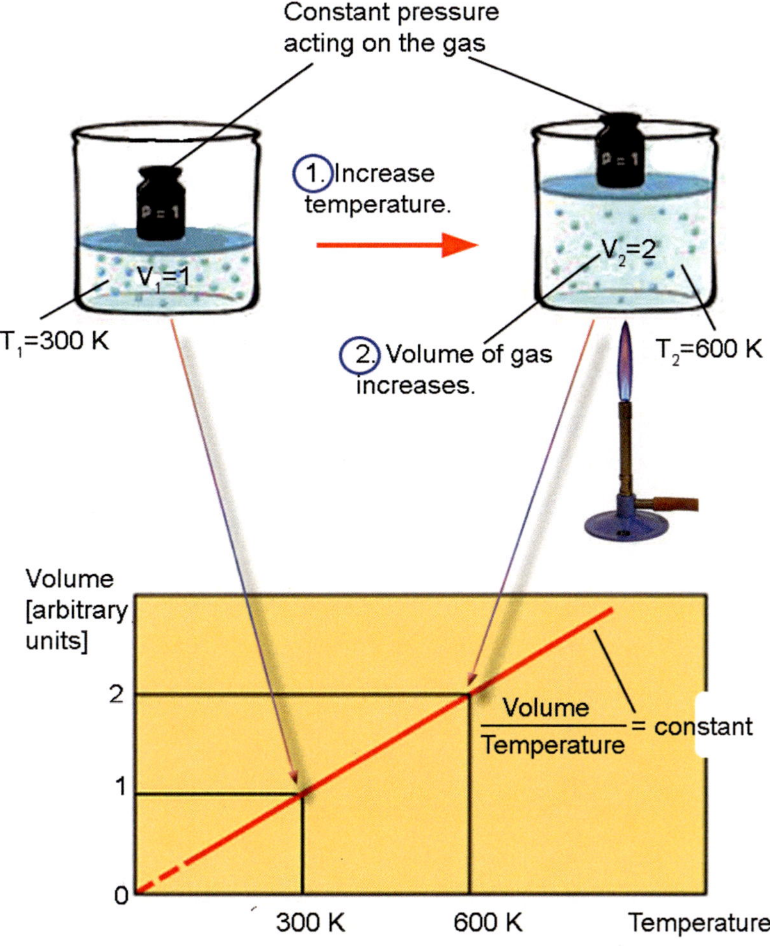

At a constant pressure, the volume V of a given amount of gas increases or decreases in direct proportion with its absolute temperature T, measured in degrees Kelvin or

$$\frac{V_1}{T_1} = \frac{V_2}{T_2} = \text{constant}$$

We can confirm Charles's law by a simple experiment. Take an inflated balloon and place it in an ice bath for a few minutes. Observe its volume and then place it in a boiling water bath. Note its dramatic increase in volume.

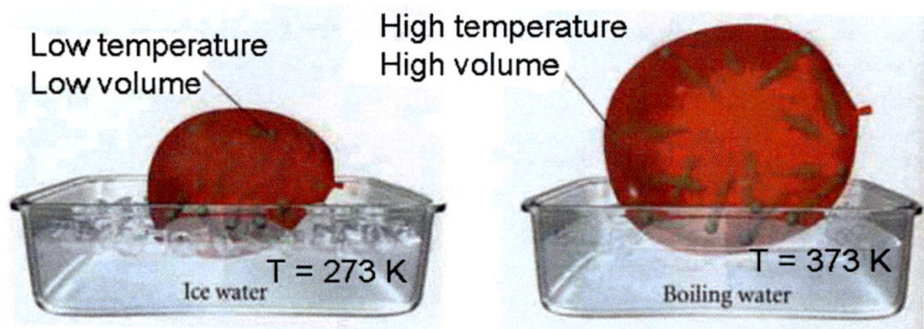

Finally, let's see what happens if we keep the volume constant and heat the gas. Molecules move faster at higher temperature and strike the container walls harder and more frequently, leading to higher pressure. *If the volume and the number of molecules are held constant, but temperature increases, pressure will increase as well.* In another words, the ratio of pressure P and temperature T is constant. This is known as *Gay-Lussac's law* (**Figure 6.25c**). In symbols,

$$\frac{P}{T} = \text{constant (for a given amount of gas held at constant volume)}$$

c. Gay-Lussac's law

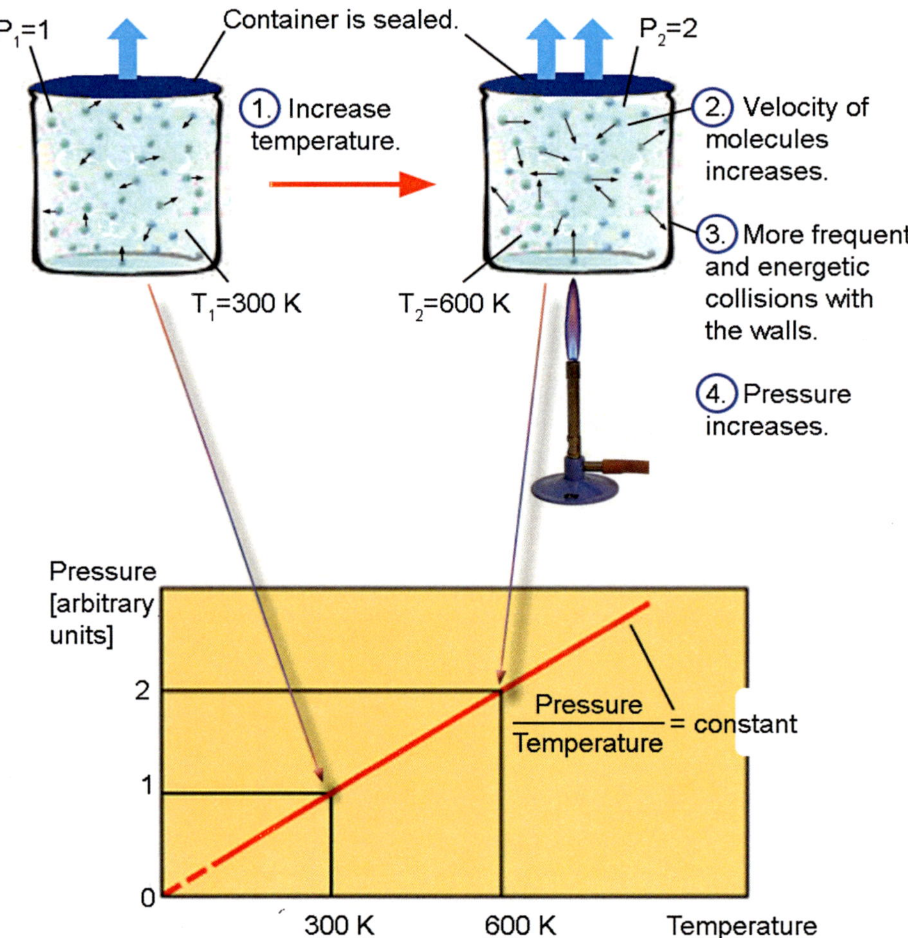

At a constant volume, the pressure P of a given amount of gas increases or decreases in direct proportion with its absolute temperature T, measured in degrees Kelvin or

$$\frac{P_1}{T_1} = \frac{P_2}{T_2} = \text{constant}$$

We can demonstrate the relationship between temperature and pressure by cooling a previously heated can partially filled with water.

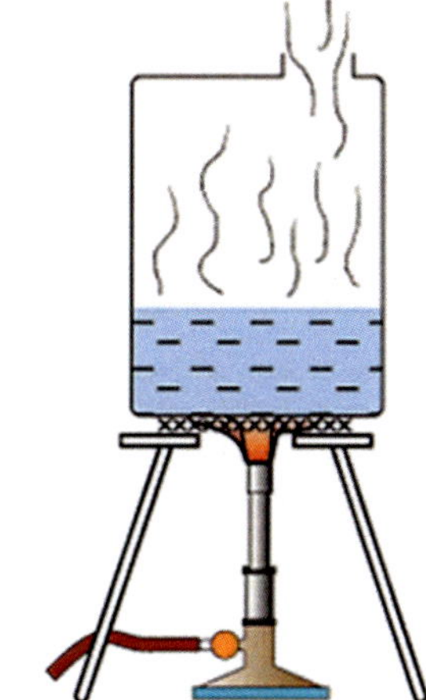

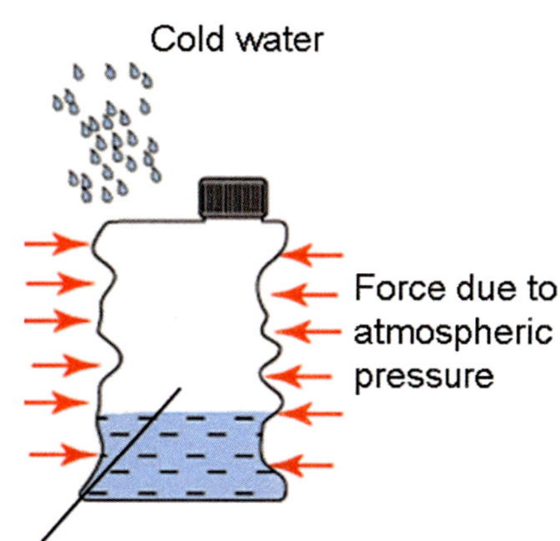

Critical thinking…
If you were to send out a weather balloon, would you want to fully inflate it before you release it and why?

We can combine the above relationships between pressure, volume, and temperature into a single relation, an **ideal gas law.** For a constant amount of gas, *the product of pressure P and volume V divided by temperature T is constant.* Or

$$\frac{PV}{T} = \text{constant}$$

Here, T is the absolute temperature *measured in Kelvin.*

ideal gas law $$\frac{\text{Pressure} \times \text{Volume}}{\text{Temperature}} = \text{constant for fixed amount of gas}.$$

More general version of the ideal gas law is given by

$$PV = NkT.$$

Here N is number of particles such as molecules, and k is the Boltzmann's constant = 1.38 10^{-23} m^2 kg s^{-2}.

Example 6.8 shows how we can apply the ideal gas law to find the temperature increase during the compression cycle of a cylinder in the engine of gasoline-based cars. We will study this process in detail in the next chapter once we examine heat engines.

Example 6.8
Compression of gas in a car cylinder

A cylinder in a car engine takes a volume of 0.05 m^3 of air into the chamber at 30° C and at atmospheric pressure. The piston then compresses the air to 0.100 times of the original volume and to 24.0 times the original pressure. What is the new temperature of the air?

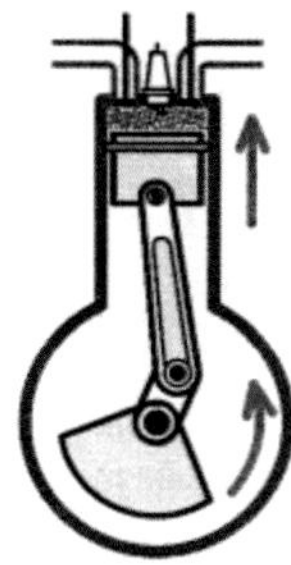

Solution:

We apply the ideal gas law

$$\frac{PV}{T} = \text{constant}$$

We select subscript 1 to describe the initial state and subscript 2 to describe the final state. Therefore,

$$\frac{P_1V_1}{T_1} = \frac{P_2V_2}{T_2} \quad \text{or} \quad \frac{(1\text{ atm})(0.05\text{ m}^3)}{303\text{ K}} = \frac{(24\text{ atm})(0.005\text{ m}^3)}{T_2}$$

By cross multiplying, we get

$0.05\ T_2 = 36$ and $T_2 = 720$ K. In Celsius, it is 720-273 = 447° C.

In the next chapter we will learn that a higher compression temperature improves the efficiency of the car.

Scuba divers understand the relationship between pressure and volume of gas well. They know it is not possible to try to breathe with a long snorkel about 1 meter (over 3 feet) under the surface of water. The muscles we use to expand and contract our lungs during breathing are not strong enough to overcome much pressure. Even just a meter beneath the surface, the pressure is great enough that we cannot expand our lungs against the water pressure to inhale a breath of air from the surface.

That's why scuba divers use a pressurized breathing gas mixture. If the breathing gas supply is delivered at the same pressure as the surrounding water pressure, the diver's lungs do not have to work against the water pressure. In other words, the pressure in the surrounding water and the pressure in the inhaled gas supply are balanced. However, gas at a higher pressure occupies a smaller volume and this could be a problem once the diver returns back to the surface.

Let's model diver's lungs as an inflatable balloon. If we fill the balloon with gas at the water's surface, and then pull the balloon underwater, the balloon will shrink in size. If you bring the balloon down to a depth of 10 meters (33 feet), its volume will be half the size it was at the surface. At 20 meters (66 feet), its volume will be one-third the size it was at the surface. See **Figure 6.26**.

Volume of a balloon at different depths Figure 6.26

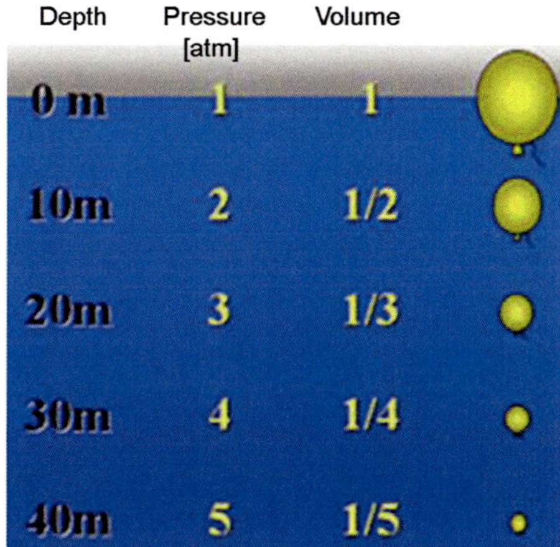

If we wanted to expand the balloon to its original size at a depth of 20 meters, we would have to fill it with three times as many gas molecules as were required at the surface. Returning the re-inflated balloon back to the surface would cause it to grow to three times its original volume. Therefore, it is not a good idea for a diver to take a full breath at 20-meter depth and hold it while ascending to the surface. The diver would run the risk of suffering from ruptured lungs.

Example 6.9 shows how party supply shops apply the ideal gas law to find the specific balloon capacity of their helium tanks that they rent.

Example 6.9
Balloon capacity of a helium tank

A tank of helium has a volume of 5 liters and a pressure of 24 atm. How many balloons can it fill if each balloon has a volume of 2 liters and requires a pressure of 1 atm?

Solution:
We apply the ideal gas law for constant temperature or PV = constant where P is pressure and V is volume. Here

$$PV = (24 \text{ atm})(5 \text{ liters}) = 120 \text{ atm liters} = (\text{Number of balloons})(2 \text{ liters})(1 \text{ atm}).$$

Number of balloons = 60.

Buoyancy of Air

We have studied buoyancy of liquids in the previous chapter. Buoyant forces occur just as readily in gases. *An object surrounded by air experiences buoyant force equal to the weight of displaced air.* This property allows some objects to float. It is similar the object's floatation condition on water's surface. Recall the floating ice cube in the previous chapter.

To gain floatation, the weight of the displaced air surrounding the object has to at least match the weight of the object, not an easy task because the density of air is only about 1.25 kg/m^3. Therefore, an object having volume of 10 m^3 (a cube about 7 feet on each side) would displace enough air to support mass of only 12.5 kg afloat. Most materials have densities which are at least thousand times, or so, bigger than density of air and an object having volume of 10 m^3 would be much too heavy to experience floatation in air.

Gases have very low densities and are an exception. A balloon filled with a gas so that it is less dense than air will float. Balloonists typically use hot air or helium. Hot air expands and becomes less dense than cool air. The buoyant force must at least equal to the weight of the balloon, gas inside the balloon, and balloon's load for the balloon to float. The air inside the balloon adds weight to the balloon, whether hot or cool. The hotter air has more kinetic energy and pushes harder on the walls of the balloon to increase its volume and therefore decrease its density. The balloon becomes less dense than the surrounding air and moves upward. At the same time, cool air spills out of the bottom, decreasing the balloon's weight. **Example 6.10** shows how hotter air changes the volume of a hot air balloon.

Example 6.10
Hot air balloon
In the photo, the flame under the balloon heats up the air which expands and increases volume of the balloon bag. The balloon becomes less dense than the surrounding air and floats. What happens to the volume of 100 m^3 of air inside the balloon's bag if its temperature is raised from 30°C to 130°C? Assume the pressure remains the same.

Solution:
We apply the ideal gas law.

$$\frac{PV}{T} = \text{constant}$$

At constant pressure

$$\frac{V}{T} = \text{constant}$$

Temperature T is measured in K so 30°C = 273 + 30 = 303 K and 130°C = 273 + 130 = 403 K. We insert volume and temperature in the above expression

$$\frac{100 \text{ m}^3}{303 \text{ K}} = \frac{Volume}{403 \text{ K}}$$

Volume = 133 m^3.

Therefore, we can significantly increase the balloon's volume and thus decrease its density by heating up the air inside of it. That's why we call them hot air balloons.

CONCEPT CHECK

1. **How** is the kinetic energy of molecules related to temperature?
2. **What** forces act between ideal gas molecules?
3. **What** happens to the volume of a balloon submerged under water to a depth of 10 m?

6.6 Fluids in motion

LEARNING OBJECTIVES

1. **Define** Bernoulli's principle.
2. **Describe** how spinning of a fast moving ball changes its trajectory.

We have studied pressure in stationary liquid in the previous chapter and learned that those same principles apply to stationary gasses. Pressure changes with depth but stays the same if you move horizontally across the stationary fluid – liquid or gas. If the fluid is moving, pressure can also change in the horizontal direction. Consider a fluid moving through a pipe that consists of two sections, each with different pipe diameter. The pressure is now different in each section of the pipe. We will find out how pressure changes along a moving fluid, define the Bernoulli's principle, and show many interesting applications ranging from curved soccer kicks to ski jumper's air lift.

Bernoulli's principle

Imagine water flowing through a pipe whose diameter suddenly becomes smaller. There is no pileup of water and we must have the same amount of water passing any given point during any given time interval, say a one second. Water must move faster through the narrower part of the pipe or a stream.

You can easily observe it by watching the speed of water flow in a river that dramatically narrows, for instance. We can quantify this relationship by

$$Av = \text{ constant}.$$

Here A is a cross section area of a pipe and v is the fluid's velocity.

What happens to pressure in the moving water as the pipe narrows? Water accelerates across the boundary and its kinetic energy increases. That means a net force does work on the water. This force is associated with the difference in pressure across the boundary. It points in the direction of faster liquid flow to the narrower part of the pipe. Therefore, the pressure in the faster moving liquid is smaller than in the slower moving liquid. See **Figure 6.27**. Daniel Bernoulli (1700-1782) first formulated it:

While the speed of a fluid increases, internal pressure in the fluid decreases.

We call it the **Bernoulli's principle**.

Bernoulli's principle When the speed of a fluid increases, internal pressure in the fluid decreases.

Bernoulli's principle Figure 6.27
Imaginary cube of water speeds up as it enters a narrower part of the pipe. As a result, pressure in the fluid in the narrower part of the pipe decreases.

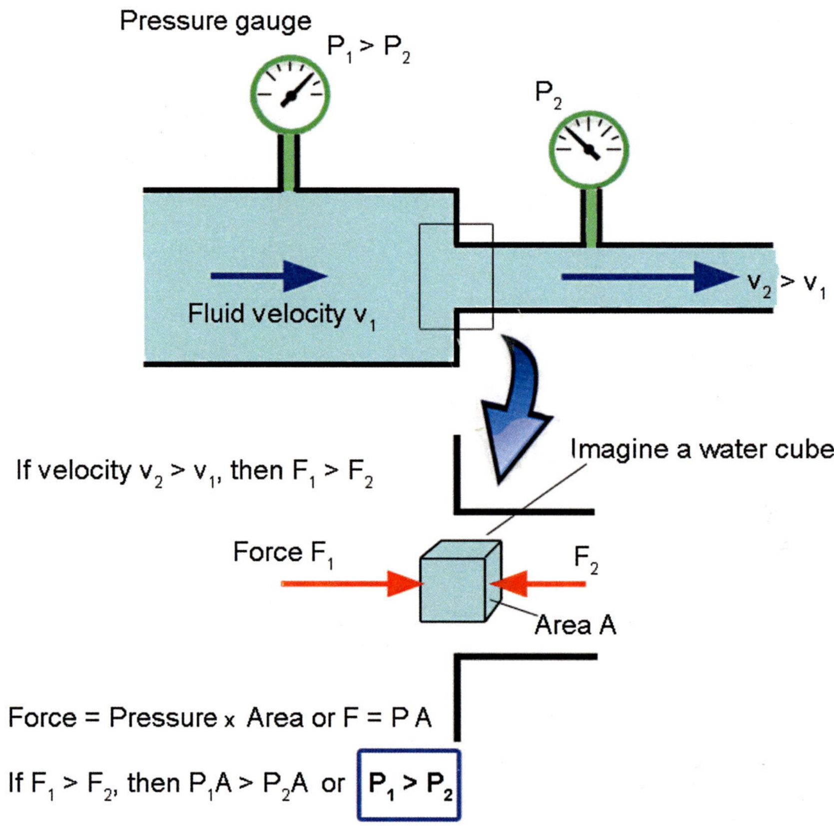

If we remove the pressure gauges and replace them with open tubes, water will rise to different heights in each tube. The higher water level h_1 in the left column needs higher pressure than the lower water level h_2 in the right column.

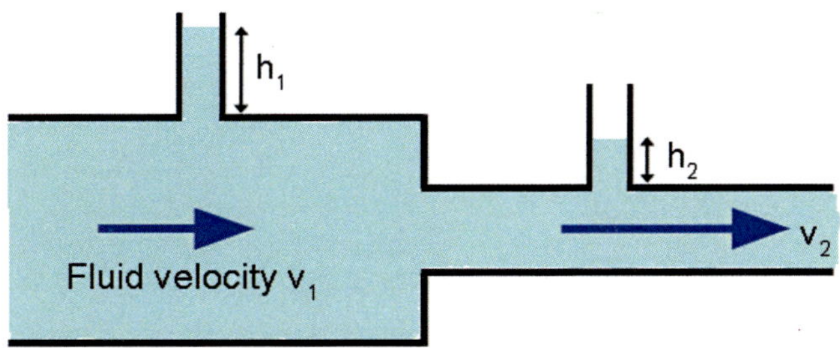

We can model the motion of a liquid or gas by drawing imaginary *streamlines* that follow the flow. Streamlines represent smooth paths of tiny bits of fluid. When they are closer together, flow speed is greater and pressure within the fluid is smaller. Bernoulli's principle applies to smooth steady flow (*laminar flow)*. When the flow becomes chaotic (*turbulent flow*), Bernoulli's principle does not work well. We assume constant density fluid here as well.

More general version of Bernoulli's principle examines the total energy density. The sum of kinetic energy per unit of volume and the potential energy per unit of volume and pressure (units are energy per volume) is constant along any streamline. Or

$$\frac{1}{2}\rho v^2 + \rho g y + P = \text{ constant}$$

Here ρ is density of the fluid, v is the fluid's velocity, y is the vertical position of any point within the fluid, P is pressure, and $g = 9.8$ m/s^2.

At first, Bernoulli's statement seems counterintuitive for you would face higher pressure if you stood in the way of faster moving fluid coming right at you. By pressure here we do not mean the pressure by the fluid but rather internal pressure within the fluid. An example of internal pressure would be a placement of an air bubble within the flow and observing it after the flow speeds up in a different location. It would increase in size because the liquid's internal pressure would be smaller. Perhaps the easiest way to demonstrate the relationship between pressure and air speed is to take a sheet of paper, hold it up to your mouth and blow across it. See **Figure 6.28**.

Fast moving air reduces pressure Figure 6.28

Hold a sheet of paper under your lips and blow across it. The paper rises as predicted by the Bernoulli's principle.

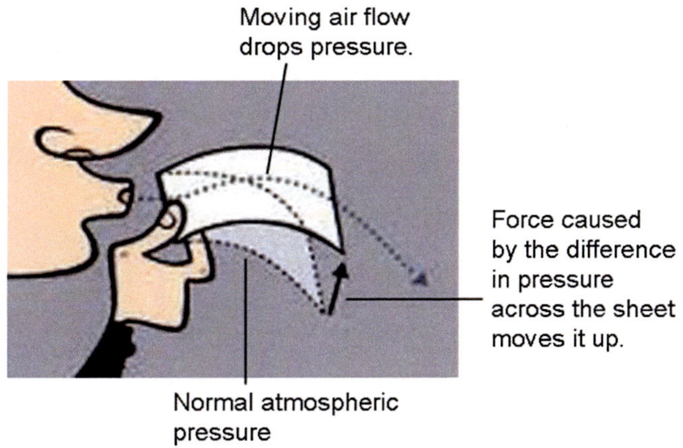

Applications of Bernoulli's principle

Many soccer fans fondly remember the free kick taken by the Brazilian Roberto Carlos in a World Cup tournament in France during summer of 1997. The ball was placed about 30 m from his opponents' goal and slightly to the right. Carlos hit the ball so far to the right that it initially cleared the wall of defenders by at least a meter and made a ball boy, who stood meters from the goal, duck his head. Then, almost magically, the ball curved to the left and entered the top right-hand corner of the goal—to the amazement of players, the goalkeeper, Fabien Barthez, and the fans alike. You can see this amazing kick on youtube.com under search words "Roberto Carlos free kick". See **Figure 6.29**.

Carlos's curved direct kick Figure 6.29

a. Roberto Carlos applied Bernoulli's principle to shoot a curve ball.

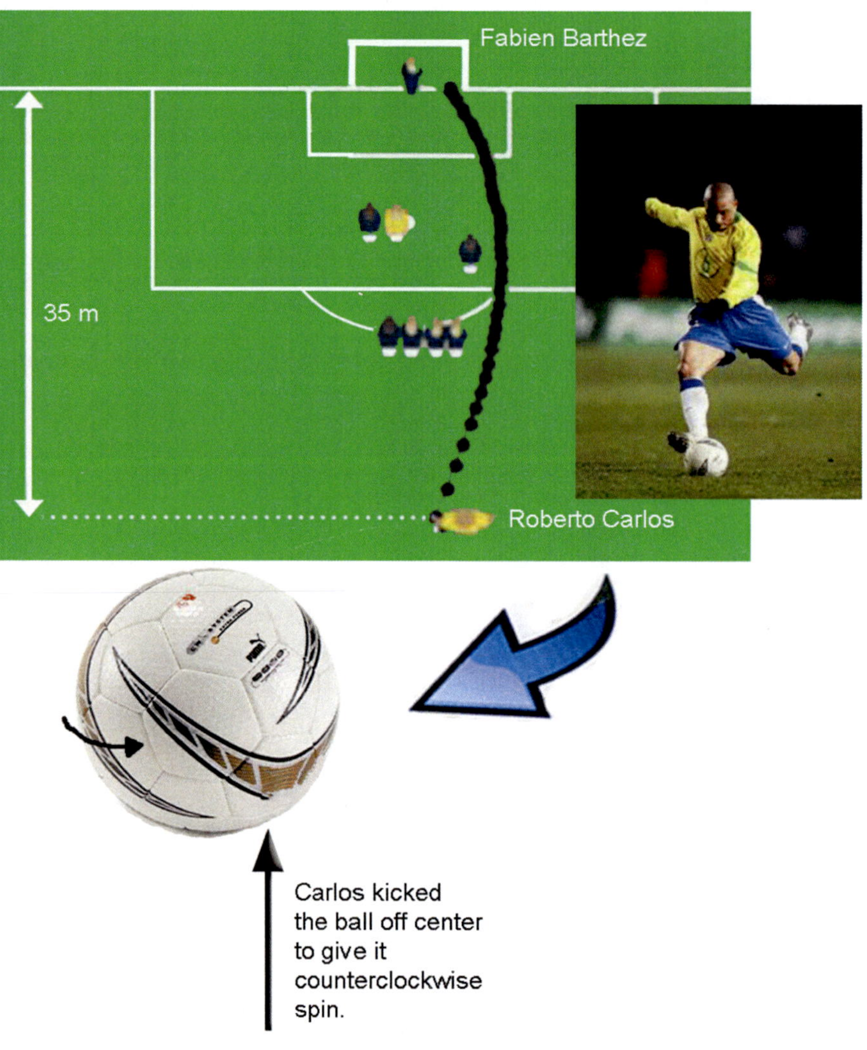

b. The moving layer of air creates crowding of streamlines on the side where the periphery of the ball is moving in the same direction as the airflow. Air pressure is lower on that side than on the opposite side and the ball curves as shown.

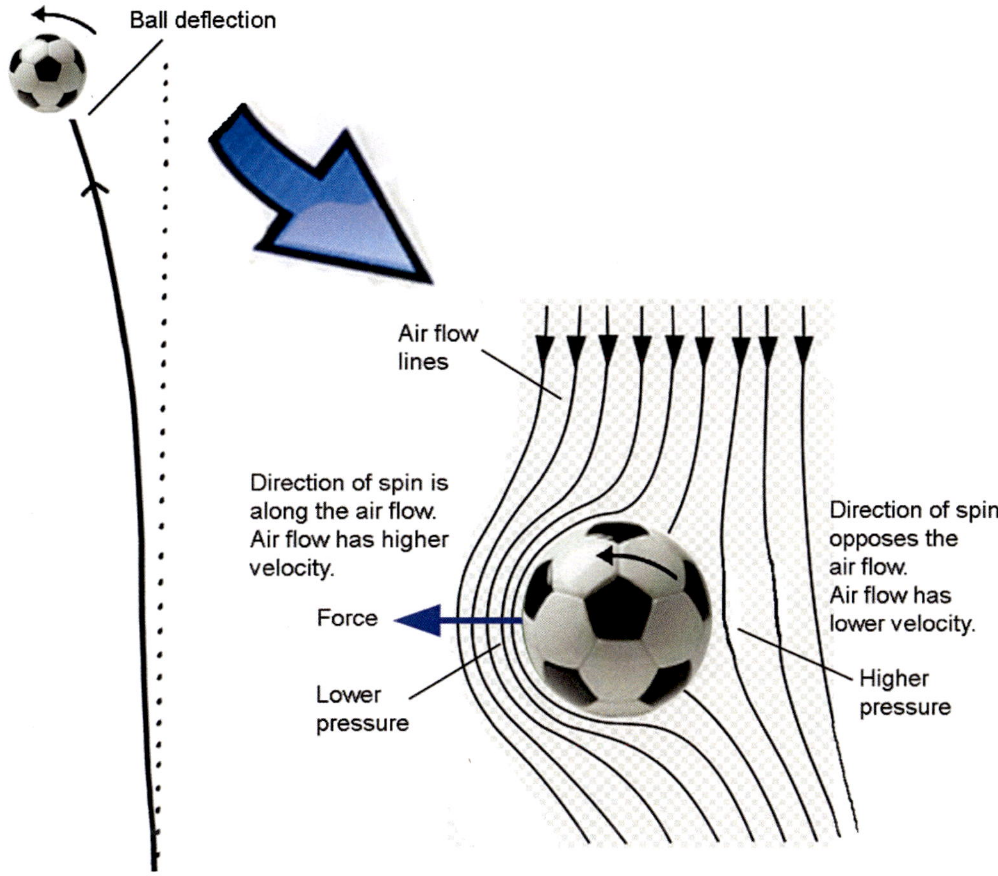

If the velocity of the ball is 25-30 m/s (about 70 mph) and that the spin is about 8-10 revolutions per second, then the lift force turns out to be about 3.5 N. A professional soccer ball has a mass of about 0.430 kg, which means that it accelerates by about 8 m/s^2. If the ball travels its 30 m trajectory in about 1 s, the lift force could make the ball deviate by as much as 4 m from its normal straight-line course.

Baseball players apply the same principles to throw a curveball. Tennis players like to place topspin on tennis balls. It curves their trajectory downward, and the balls reach the ground sooner. Pro ping-pong players smash their balls hard, yet they never seemed to land past the table. The spin on the ping-pong makes it drop faster.

Bernoulli's effect can explain the lift force on a roof when a fast moving wind blows across the roof. The wind gains speed as it reaches the roof's peak. Streamlines there get more crowded and pressure along the streamlines becomes smaller. This is similar to the crowding of streamlines as they move through a constricted region.

At the same time, the pressure under the roof is the higher atmospheric pressure. The difference between the pressures over a large area of the roof causes an upward force on the roof that can be significant. That is why we put vents in the roofs of houses to help equalize the pressure. See **Figure 6.30**. Understanding the lift force on the roof, we can examine the lift force on an airplane wing. The airplane wing is shaped and oriented so that during flight, air travels faster over its top surface than under its bottom surface. A higher pressure underneath the wing than above the wing provides an upward lift force. See **Figure 6.31**.

Lift force on a roof Figure 6.30

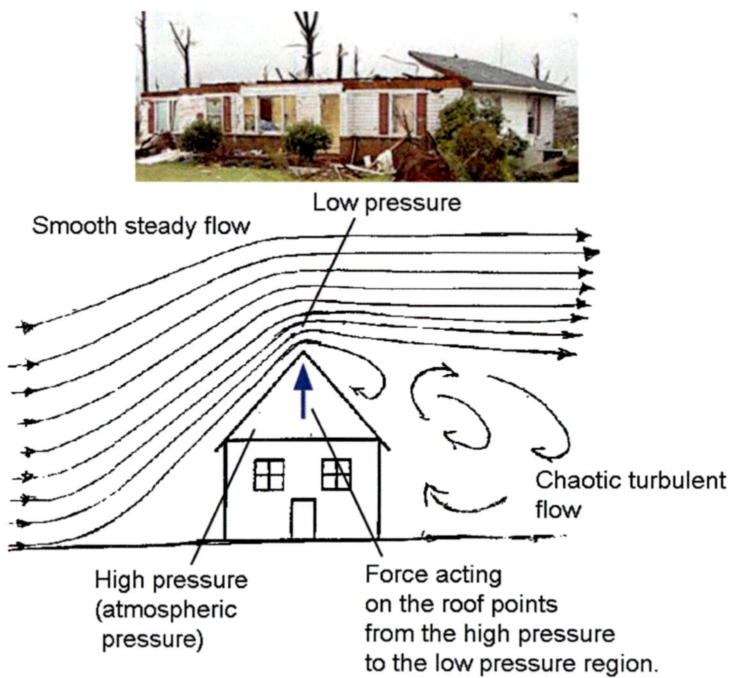

Lift of an airplane wing Figure 6.31

Lift force acting on an airplane wing is the result of pressure difference between the top and bottom of the wing. The net upward pressure on the wing multiplied by the surface area of the wing gives the net lift force.

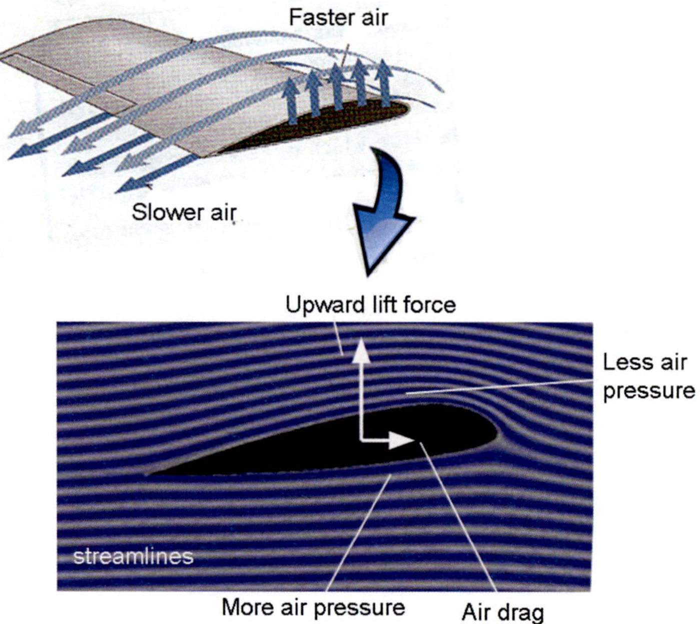

A pilot can change the lift force and plane's altitude by tilting the wings. The angle at which the wings approach the air stream is called an *angle of attack*. By increasing the angle of attack angle, the speed difference between the air flowing over and under the wing increases, and results in an increase of the lift force. Thus, the pilot rotates the nose of the plane upwards to accelerate upward, and downwards to accelerate downward. By selecting the right angle of attack a skilled pilot would be able to fly upside down although the situation would be less stable. This trick is easier when the plane's wing has the same curvature on top and bottom.

There is a limit to increasing the angle of attack. Beyond a certain angle, the airflow over the top of the wing separates from its surface and the wing stalls. A turbulent air pocket forms above the wing, reducing the wing's lift. If the lift force drops below the weight of the airplane, the airplane begins to lose altitude. Pilots need to fly at higher speeds to maintain higher lift forces. For a commercial jet, this speed is about 220 km/h. Commercial jets can also change the shape of the wings to vary the lift force, which is proportional to the wings surface area. Slats move forward and down the leading edges while flaps move back and down from the trailing edges of the wings. **What a Physicist Sees** takes the wing concept further and examines the lift force on a ski jumper.

What's a Physicist Sees?
V-shape ski jumping

A physicist sees the ski jumper, while pressing his body parallel to the skis, acting like an airplane wing. The air moves faster over the upper surface of his body than underneath his skis. This gives him a lift force. However, the lift force is smaller than the force of gravity, so the jumper gradually comes down. In late 70s, some skiers experimented with positioning their skis in a V-shape during flight. They discovered they could fly further.

This was because the V-shape of skis increased their "sail" area, and that allowed for a larger lift force. Now everybody uses the V-shape jumping style. Any aircraft engineer knows that the lift is proportional to the product of airspeed-squared and the wing area.

Critical thinking…

1. Can you think of a water sport that uses vertical "wing" to produce a sideways force? The force is sideways yet the overall motion is forward; why?
2. Can you think of an application of an inverted "wing"? How would it work?

There are more interesting examples. See **Figure 6.32**.

Artery constriction Figure 6.32
The constriction in the blood vessel speeds up blood going through it. The lower pressure causes the vessel to close, stopping the flow. Without flow, there is no Bernoulli's effect and blood pressure causes it to re-open. The process repeats.

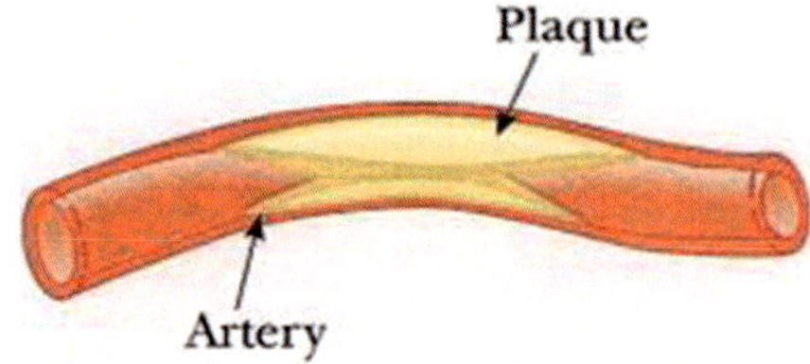

Viscous Flow

Bernoulli's equation states that when a fluid flows steadily through a long, narrow, horizontal pipe of constant cross section, the pressure along the pipe is constant. In practice, we observe a pressure drop as we move along the direction of flow. This means that pressure difference is required to push the fluid through the pipe. There are drag forces between the different layers of the fluid and the layer of fluid in contact with the pipe. We called them **viscous forces.**

We define the pressure drop between two pints along the pipe is

$$\Delta P = P_1 - P_2 = AvR$$

where Av is the volume flow rate and the proportionality constant R is the resistance to the flow. It depends on the length L of the pipe, the radius r, and the viscosity of the fluid η as

$$R = \frac{8\eta L}{\pi r^4}$$

This means that the pressure drop can be written as

$$\Delta P = \frac{8\eta L}{\pi r^4} Av$$

This equation is known as **Poiseuille's law.**

Note the inverse r^4 dependence of the pressure drop. If the radius is halved, the pressure drop across the tube increases by a factor of 16. Think of the pressure drop increase in your arteries as they get partially clogged by cholesterol.

CONCEPT CHECK

1. **What** happens to air pressure as you blow over the top of a sheet of paper while holding it under your mouth?
2. **Why** do table tennis players put topspin on the ping-pong balls?

Summary

6.1 Pressure in Liquids and Pascal's Principle

- Liquids and gases have many common properties. Because they both flow, they are grouped together under the common name of **fluids**. Liquids are not compressible, but gases are. We can say about **static fluids** (fluids at rest) that
  1. At any point in a fluid, pressure acts equally in all directions.
  2. Pressure increases with depth.
  3. Pressure does not change as you move horizontally within and across the fluid.
- The pressure anywhere within a fluid depends only on the depth and the density of the fluid, and not on the volume of the vessel, as shown here.
- **Pascal's Principle** – Pressure applied at any point in an enclosed liquid at rest is transmitted undiminished to all other points within the liquid, as shown here.
- **Hydraulic lifts** are based on this principle, enabling them to lift heavy objects.

6.2 Pressure at Different Depths

- Pressure P in a static fluid at depth h and density of the fluid D is

  $$P = Dhg$$

  where g is the acceleration due to gravity. The total pressure inside a fluid that is open to the atmosphere is obtained by combining this pressure with atmospheric pressure, or

  $$P_{\text{abs}} = P_{\text{atm}} + Dhg$$
- A device that measures atmospheric pressure is called a **barometer**, as shown here.
- **Gauge pressure** is a pressure reading without taking into account atmospheric pressure. Tire pressure and blood pressure readings are examples of gauge pressures.

6.3 Archimedes' Principle and Flotation

- An object partially or wholly immersed in a liquid is acted upon by an upward buoyant force equal to the weight of the liquid it displaces.
  1. A buoyant force acts on an object immersed in a liquid, as shown below, because the pressure beneath the object is larger than the pressure above. The resultant force on the object is upward, opposing the force of gravity.
  2. The apparent loss in weight of an object immersed in a liquid is due to the buoyant force.
  3. When an object is partially or fully submerged, the buoyant force, or apparent loss in weight, is equal to the weight of the fluid displaced. This is **Archimedes' principle**.
  4. The apparent weight equals the actual weight minus the buoyant force.
  5. A floating object displaces its own weight in liquid form. The buoyant force is equal to the weight of the object.
  6. A floating object has a density that is less than the density of the liquid in which it is floating.
- Water's surface experiences tension, called **surface tension**. For tiny objects, the surface acts like an elastic membrane and small objects can float on it.

6.4 Earth's Atmosphere

- Atmosphere forms a thin layer of gases around the Earth. More than 99% of the total mass of the air is below 40 km (25 miles). The density of air steadily decreases with altitude. At 40 km it is less than 1% of its value at sea level.
- We live under the weight of an ocean of air. This air exerts pressure on us just like water exerts pressure on a diver. We call it the **atmospheric pressure**. At sea level this pressure is about 101,000 Pa.

6.5 Ideal Gas and its Properties

- A gas molecule undergoes up to 2 billion collisions each second. **Temperature** is a measure of the average kinetic energy of the molecules of a substance. We use Celsius, Fahrenheit, and Kelvin scales to measure temperature.

- **Absolute temperature scale** uses Kelvin degrees and has 0 at –273.15°C. One converts to Kelvin degrees by adding 273.15 to Celsius degrees.
- Gas is called **ideal gas** if its molecules are so far apart that they exert forces on each other only when they collide. The molecules travel freely in a random straight-line motion until they hit each other or hit the walls of their container.
- Reducing the volume of a gas increases its pressure if the temperature and the amount of gas are constant.
- Volume of gas is directly proportional to its temperature at constant pressure.
- If the volume and the number of molecules are held constant, but temperature increases, pressure will increase as well.
- **Ideal gas law**: $\frac{\text{Pressure} \times \text{Volume}}{\text{Temperature}} = \text{constant for fixed amount of gas}\cdot$

6.6 Fluids in motion

- **Bernoulli's principle**: While the speed of a fluid increases, internal pressure in the fluid decreases.
- We can model the motion of a liquid or gas by drawing imaginary *streamlines* that follow the flow. Streamlines represent smooth paths of tiny bits of fluid. When they are closer together, flow speed is greater and pressure within the fluid is smaller. Bernoulli's principle applies to smooth steady flow (*laminar flow).*

Key Terms

fluids
Pascal's principle
hydraulic lift
pressure at depth h
barometer
gauge pressure
buoyant force
Archimedes' principle
surface tension
Atmospheric pressure
Temperature
Absolute temperature scale
Ideal gas
Ideal gas law
Bernoulli's principle

Critical Thinking Questions

1. Imagine you take two cans, one capped and one uncapped, attach weights to both, and throw them into the ocean. What will happen to each can?
2. A metal cube sinks easily, but large metal ships float. Explain.
3. Which weighs more: a pound of feathers or a pound of iron? (*Hint:* "They are equal" is not the right answer.)
4. The Arctic ice pack is made of frozen seawater. However, Antarctic icebergs are blocks of fresh-water ice from the continent that are floating in the salt-water sea. When the Antarctic ice melts, what may happen to the water level?
5. If you place a floating object from fresh water into salt water, what happens to its water line? Explain.
6. To measure blood pressure, we place the blood-pressure cuff on the arm at the same height as the heart. What physics principle do we apply?

7. Community water towers, such as this one, have shapes where most of the water sits in the upper third of the tower. Why is that better than having a regular cylindrical tower with normal vertical sides running from the ground to the top?
8. Giraffes have a systolic blood pressure of about 350 mm of Hg, and their hearts have a mass of about 12 kg. Explain why this high blood pressure is necessary.
9. If the diver in the photo below changes his orientation, will the pressure on his mask change? What if he changes his depth?
10. How could you determine if a piece of jewelry you have is made out of solid gold? Would it help to determine its volume?
11. If liquid pressure were the same at all depths, what would happen to buoyant force?
12. Is it correct to say that heavy objects sink and light objects float? Why?
13. You have two cubes of the same size, one made out of aluminum and one made out of wood. You place them in a glass of water. The wooden cube floats and the aluminum cube sinks. Which one, if either, experiences the greater buoyant force?
14. You tie a piece of iron to a block of wood floating on water in a container. Then you turn the block over. What happens to the water line of the wooden block? Will the water level change? Explain.
15. If you place a small needle on its side in water, it floats. Now drop the needle point-first in the water; it sinks. Why?
16. Place an ice cube into a graduated cylinder filled with water and mark its level. What happens to the water level when the ice cube melts? Explain. Now assume that the floating ice cube has air bubbles. After it melts, what happens to the water level in the glass? What if grains of sand froze in the ice cube? How would the water level change?
17. How can race car engineers apply Bernoulli's principle to improve the car's traction?
18. Why is it dangerous to stand near the edge of a platform when a train goes by at high speed?
19. How does Bernoulli's principle work in sailing?
20. What happens if you blow air between two burning candles standing side by side? Will the flames of the candles tilt toward or away from each other, if any, and why?
21. What would happen to the Earth's atmosphere if there were no gravity?
22. Did you notice that a stream of water flowing uniformly from a faucet becomes thinner as it travels downward? Can you explain that?
23. Explain how a suction cup holds on to a surface.
24. Why is it not a good idea to hold your breath while rising during scuba diving?
25. What happens to a diver's lungs as she goes deeper?
26. People often say "hot air rises" when explaining the operation of a hot air balloon. However, that does not explain why the basket and the load lift as well. Can you give a better and more thorough explanation?

Exercises

1. The Goodyear blimp uses about 6000 m^3 of helium. If air has a density of 1.25 kg/m^3 and helium has a density of 0.18 kg/m^3, find the buoyant force on the blimp.
2. What happens to the volume 100 cm^3 of air if its temperature is raised from 20°C to 120°C? Assume the pressure remains the same.
3. What happens to the volume of the hot air in the previous exercise when the pressure drops in half, which occurs at an altitude of 5,500 m?
4. The gas in a cylinder of a diesel engine is at a temperature of 47°C and a pressure of 1 atm. After compression, the temperature is 700°C and the pressure is 48.5 atm. What is the ratio of the initial to the final volume? This is also known as the compression ratio of a diesel engine.
5. We have a gas in a cylinder with a movable piston. We record its initial pressure, volume, and temperature. We heat the gas to a new temperature that is four times the original temperature as measured in Kelvin. If the volume increases to three times the original volume, what is the new pressure?

6. A tank of helium has a volume of 4 liters and a pressure of 80 atm. How many balloons can it fill if each balloon has a volume of 2 liters and requires a pressure of 1 atm?
7. An ideal gas is confined within a closed cylinder at atmospheric pressure (1.013×10^5 Pa) by a piston. The piston moves until the volume of the gas is reduced to one-sixth of the initial volume. What is the final pressure of the gas when its temperature returns to its initial value?
8. Calculate the net force on the front door of a closed house if the outside pressure suddenly dropped 10 percent. The dimensions of the door are 7 feet times 3 feet. This pressure drop could occur in a tornado.
9. Scuba divers use double tanks that hold 71 ft^3 of sea level air each. What is the weight of the air if the air's density is 0.08 lb/ft^3?
10. A free diver descended 100 m deep on a single breath. What was the ratio of his lung volume at that depth to his lung volume when he returned to the surface?
11. What is the pressure at a depth of 300 m in seawater?
12. If the atmospheric pressure, 101,000 Pa, were to support a column of water, how high would the water rise? See the figure below. Assume that no pressure acts at the top of the column. Answer the same question for mercury. The density of water is 1000 kg/m^3; the density of mercury is 13,600 kg/m^3.
13. A plastic bag contains a glucose solution. If the gauge pressure in the artery is 13,300 Pa, what must be the minimum height of the bag relative to the artery to allow the glucose into the artery?
14. A life jacket has volume of 0.05 m^3. What is the buoyant force exerted on it by water? The density of water is 1000 kg/m^3.
15. A sphere of aluminum has a volume of 0.01 m^3. In air it weighs 265 N. What is the apparent weight of the sphere when it is completely submerged under water?
16. A man has a surface area of 2 m^2. Find the total force on his body due to atmospheric pressure. If this force were the weight of an object, what would be its equivalent mass?
17. A driver's foot applies a force of 100 N to a piston of area 25 cm^2 in a master cylinder shown in the figure below. The output piston in the brake cylinder has an area of 50 cm^2. What is the pressure in the fluid? What is the force exerted on the piston in the brake cylinder by the fluid?
18. Blaise Pascal arranged a spectacular demonstration, illustrated below, in a Parisian garden. A slender tube, over 12 m long and 0.006 m in diameter, ran up from a small hole in the side of a strong oak barrel, reinforced with iron straps, all the way up to a balcony of a tall building. Pascal poured a little bit of water into the tube from the top of the balcony and burst the barrel wide open, stunning his audience. Find the pressure at the bottom of the tube, and then determine the force that was generated there. Keep in mind that the tube was open to the atmosphere at the top.
19. A cube of surface area 6 m^2 is submerged in water, as shown here. Pressure on the top face is 121,000 Pa and the pressure on the bottom face is 130,800 Pa. What is the buoyant force acting on the cube?
20. A water tower is bulb-shaped such that the bottom of the bulb is 30 m above the ground. What is the lowest maintenance pressure in the tower?
21. If the density of a floating object increases by a factor of three but its mass stays the same, how much does the buoyant force change?
22. How much volume would 1,000 kg of iron need to occupy to float? The density of iron is 7,800 kg/m^3.

Chapter 7
Heat

Mature sequoias survive all but the very hottest fires because their thick, fibrous bark with many air pockets insulates the trees from lethal heat. The bark holds very little sap or pitch, so it is not very flammable. In this chapter, we will learn about how materials conduct heat and why some materials, such as the bark of a sequoia tree, are good thermal insulators while others, such as metals, are not.

All matter contains hidden thermal energy, hidden not only in the motion of its molecules, atoms, and electrons, but also in the chemical bonds between them. We are aware of this thermal energy upon physical contact, as we immediately know if something is hot, cold, or in between. For instance, when we touch a hot pot, thermal energy flows from the pot to our fingers. This flow of thermal energy is called heat.

In this chapter, we will examine heat, the capacity of different objects to hold and transfer heat, and heat's role in the thermal expansion of objects. In addition to moving from one place to another, heat can transform matter from a solid to liquid state and then from a liquid to gaseous state. We will show how the addition or subtraction of thermal energy determines the different states of matter and apply this knowledge to several examples, such as evaporative cooling.

7.1 Heat and Thermal Properties of Matter

LEARNING OBJECTIVES

1. **Describe** heat.
2. **Define** specific heat.
3. **Explain** thermal expansion.

The concept of heat puzzled scientists back in the 18th century. Many assumed that matter contains some invisible fluid that flows from hotter objects to colder ones. This hypothesis was disproved once scientists realized that matter contains molecular kinetic energy and potential energy, but no invisible fluid. Our current view is that heat is energy *flowing* between two objects as a result of their temperature difference.

Nature of heat

When you touch a hot cup of tea, energy enters your hand because the cup is warmer than your hand. When you touch an ice cube, energy passes out of your hand and into the ice. The energy that transfers from the hotter object to the colder object is called **heat**.

heat Thermal energy that flows from a substance at a higher temperature to a substance at a lower temperature.

Matter contains **thermal** (or internal) **energy** that is not generally visible to the eye. This thermal energy consists of the kinetic energy of atoms and molecules and also their potential energy. The kinetic energy can take on three forms: kinetic energy of translation, rotation, and vibration (**Figure 7.1**). Atoms and molecules in liquids and solids stretch apart when they gain energy and store this energy as potential energy in their spring-like bonds.

We need to distinguish between heat and thermal energy. Matter contains thermal energy and not heat. Heat is a transfer of thermal energy.

thermal energy Sum of all molecular energies, including kinetic and potential energies, internal to a substance.

Thermal energy of matter Figure 7.1
Thermal energy includes molecular translational, vibrational, and rotational motion. It can also exist as stored potential energy, a result of any bonding forces between atoms and molecules.

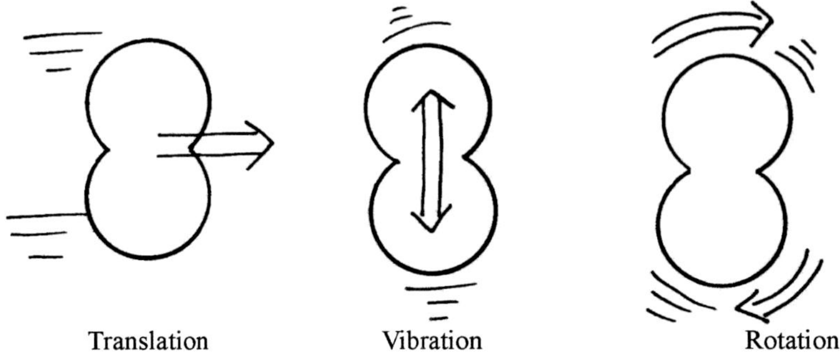

We saw in Chapter 6 that temperature is proportional to the average translational kinetic energy of the atoms or molecules. Therefore, any energy that does not increase the molecule's translational kinetic energy does not increase the molecule's temperature. For instance, when we add heat to a gas, only some of the added energy goes to increase the average translational kinetic energy of the molecules and, therefore, their temperature. Some energy may go into rotational kinetic energy of the molecules and some energy may go into vibrational motion. Those contributions do not increase the temperature.

Heat and temperature are two distinct concepts. Adding equal amounts of heat to different amounts of material produces different temperature changes. Thermal energy varies with both the mass and the temperature of an object. A teapot full of tea has more thermal energy than a cup of tea. They are at the same temperature, so the average kinetic energy of the particles is the same in both containers. However, the teapot contains more molecules and thus has more thermal energy.

Heat is not a static quantity. If we bring two objects at different temperatures into thermal contact with each other, energy flows between them, with energy flowing from the hotter object to the colder. *The heat flow from hot to cold objects is spontaneous* and is proportional to the mass of the hotter substance and the temperature difference between the hot and cold substances.

We can understand the heat flow by examining what is happening on a molecular level. The molecules of the hotter substance have a higher average kinetic energy. These energetic molecules lose some of their kinetic energy when they collide with the slower molecules of the colder substance. The average molecular kinetic energy decreases within the hotter substance and increases within the colder substance until they become equal. The hotter substance's temperature drops and the colder substance's temperature rises. When the flow of energy stops, the two substances have reached an *equilibrium temperature.*

Many different units are used in measuring heat or thermal energy. Heat is energy and therefore the metric units are joules. However, we often use other units such as *calories* or *Calories*, for instance. A calorie (c) is defined as the amount of heat required to raise the temperature of 1 gram of water by 1 degree Celsius. The energy content of food items is labeled by Calories (C). One Calorie is the amount of heat required to raise 1 kilogram of water by 1 degree Celsius and is equivalent to 1000 calories. One calorie is equivalent to 4.184 joules.

Specific heat

You might have noticed that sand (which is mostly silicon) and metal both heat up (or cool down) more easily than water. For instance, heating an empty copper pot will raise its temperature much faster than the temperature of water inside another pot. Different materials heat up at different rates because they have different capacities for storing added energy. The same amount of heat will raise the temperature of 1 gram of sand by a larger amount than 1 gram of water. We often use wooden spoons to stir boiling liquids because wood does not warm up easily. It takes more than three times as much energy to heat up a wooden spoon by 1°C than a stainless steel spoon of the same mass.

We define **specific heat** to tell us how easily a given material heats up. Specific heat is the amount of thermal energy needed to heat up a given amount of the material by one Celsius degree. This number also tells us how much energy needs to leave the material to cool down by one Celsius degree. See **Table 7.1**.

specific heat Amount of heat per unit of mass needed to raise the temperature of a substance by 1 Celsius degree.

Specific heat of selected materials, in J/kg°C Table 7.1

Solids	
Human body, average	3,470
Ice	2,100
Wood	1,800
Aluminum	900
Marble	860
Glass	840
Stainless steel (304)	500
Copper	386
Silver	233
Gold	126
Lead	128
Liquids and gases	
Water	4,186
Steam	2,100
Air (100°C)	1,000
Mercury	140

A larger specific heat means that it is harder to either warm up or cool down a given material. For instance, it takes 4.18 joules (1 calorie) of energy to heat up 1 gram of water by 1 degree Celsius. On the other hand, it takes only 0.126 joules to warm up one gram of gold, platinum, or lead, about 30 times less energy than water. Water must absorb more heat than most other substances to raise its temperature by one degree; equivalently, it must give off more heat to drop its temperature by one degree. This property is useful for regulation of body temperature in humans and animals, who must maintain a relatively constant body temperature. Our blood's resistance to a change in temperature helps us cope better with large temperature shifts.

Water's capacity to hold a large amount of heat plays an important role in Earth's climate. During warm weather, sunlight provides heat to raise ocean temperatures. However, the oceans have a huge mass of water with a high specific heat, and therefore it takes a long time for the water to warm up or cool off. Land, on the other hand, has a lower specific heat, and only a thin surface layer is heated by the Sun. Therefore, land warms up and cools off more easily than oceans.

Because of its large specific heat capacity, an ocean does not vary much in temperature from summer to winter. Land bordering the oceans benefits from water's capacity to absorb large amounts of heat during the summer months, and later it benefits again when water releases large amounts of heat during the winter months.

For instance, the west coast of the United States, aided by air flow from the Pacific Ocean into the coastal regions, has moderate temperatures year round. During the winter, the ocean's water is warmer than the air that moves over it. It warms the air, which dumps this thermal energy into the coastal regions. In summer the water is colder than the air, and it cools the air that moves toward the coast. The cool air keeps the summer temperatures moderate along the west coast. In the Midwest, on the other hand, seasonal temperature variations are large because land quickly heats up during the summer and quickly cools off during the winter.

Along the U.S. East Coast, the situation is different. Ocean current (the Gulf Stream) in the Atlantic Ocean carries warm water from the Caribbean northward along the coast and then on toward Europe. The current cools and releases its energy along the way. This current is warmer than along the U.S. West Coast, and it causes the East Coast to have higher temperatures during the summers. This same current is also responsible for Europe's warmer climate. Without it, Europe's temperatures would resemble those of northeastern Canada, since their latitudes are similar.

Thermal equilibrium

Understanding specific heat enables us to analyze the heat flow from one object to another. We already know that heat flow increases with a larger temperature difference between the hot and cold objects and with a larger mass of the heat-giving object. Heat flow also depends on the object's specific heat. A larger specific heat of an object means that the object needs to absorb or emit more thermal energy to change the temperature by 1 degree.

We combine these observations to define heat flow Q as
Heat added or subtracted Q = Specific heat c × mass m × change in temperature ΔT
In symbols,

$$Q = mc\Delta T$$

For instance, because the specific heat of water is 4.18 J/g °C (1 cal/g °C), heating 1 gram of water by 100°C requires us to add heat of

$$Q = (1\text{ g})(4.18\text{ J/g °C})(100\text{°C}) = 418\text{ J.}$$

Example 7.1 shows how we can measure an exchange of heat between two substances in an insulated vessel, called a calorimeter, and use that information to find the specific heat of one of the substances.

Example 7.1
Finding the specific heat of aluminum

A calorimeter is a double-walled container designed to keep any heat exchange within itself. A student drops a hot piece of aluminum, whose specific heat she wants to find, into cold water in the calorimeter and records the final equilibrium temperature of both. She also records the mass of each substance prior to the measurement. Her data log shows that the mass of aluminum is 0.100 kg, the mass of water is 0.200 kg, the temperature change of the aluminum is 37.2 °C, and the temperature change of water is 4.0 °C. She assumes that heat exchanged with the calorimeter is negligible. What specific heat of aluminum did she measure?

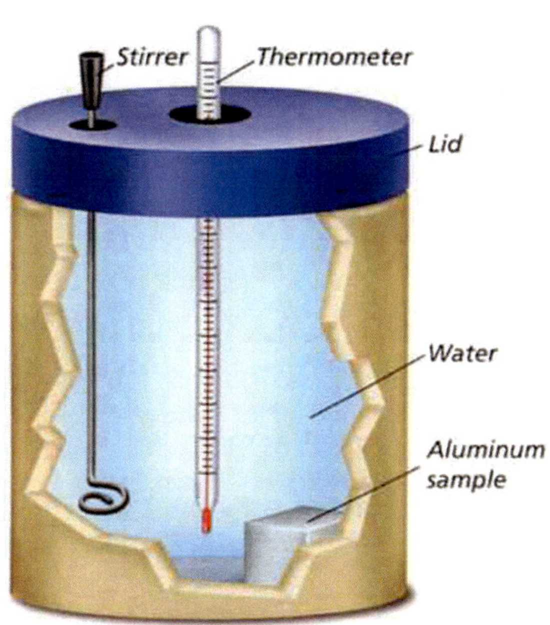

Solution: Since initially the aluminum is hotter than the water, it emits heat and water absorbs the heat. In the absence of any heat losses to the environment, we can apply conservation of energy,

Heat emitted by aluminum = Heat absorbed by water

Heat emitted by aluminum = specific heat of aluminum c_{Al} × mass of aluminum m_{Al} × change in temperature of aluminum ΔT_{Al} or

$Q_{Al} = (mc\Delta T)_{Al}$

Heat absorbed by water = specific heat of water c_w × mass of water m_w × change in temperature of water ΔT_w or

$Q_w = (mc\Delta T)_w$

Therefore,

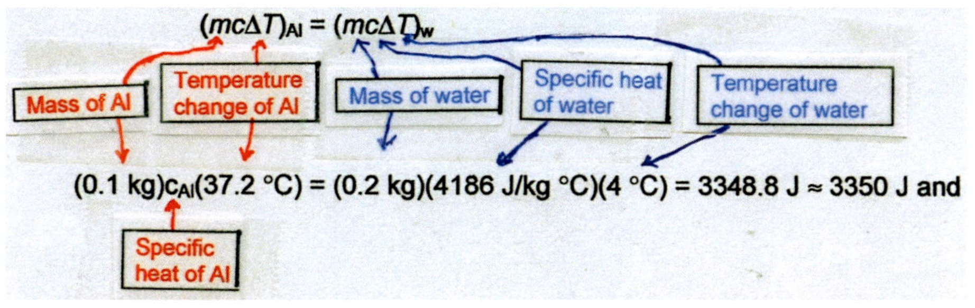

c_{Al} = 900 J/kg °C.
It takes 900 joules of thermal energy to warm up 1 kg of aluminum by 1 °C.

If we place two objects with different initial temperatures in thermal contact, they may not experience the same changes in temperature as they reach the equilibrium temperature, even if they have equal mass. The reason is that they may have different specific heats. An object with a smaller specific heat experiences a larger temperature change in this situation. A hot metal object placed in water that has the same mass as the metal will undergo a larger change in temperature than the water because the metal has a smaller specific heat.

Thermal expansion

Have you ever seen someone running hot water over a jar lid to loosen it? Then you may already be familiar with thermal expansion. A jar lid expands a little when heated, which is why running hot water over it loosens it. Gases, liquids, and solids expand when they are heated and contract when they are cooled. There are only a few exceptions. When we warm a liquid or a solid, the molecules move farther and faster on average, and their bonds stretch. This causes a slight increase in volume.

Hard rubber exposed to temperature changes expands or contracts more than most other materials. Materials with high melting temperatures show stronger bonds between atoms and expand or contract less. For instance, diamonds have a very high melting temperature and expand the least. Volume expansion is larger in liquids than solids because the bonds between molecules are weaker in liquids. Thermal expansion of liquid mercury is the basis of many thermometers.

When building roads, bridges, or railroads, engineers leave small gaps between different segments of structures, allowing for thermal expansion. These gaps are called expansion joints. Without expansion joints, buckling and warping may occur. Buckling is a failure mode in pavement materials, such as concrete. Radiant heat from the sun is absorbed in the road surface, causing it to expand and forcing adjacent pieces to push against each other. If the stress is great enough, the pavement can lift up and crack without warning. Similarly, rail tracks also expand when heated, and can fail by buckling, a phenomenon called sun kink. It is more common for rails to move laterally, often pulling the underlying railroad ties along. See **Figures 7.2** and **7.3**.

Expansion joints Figure 7.2
Interlocking expansion joints are placed between segments of roads and bridges. The gaps allow a bridge and a road to expand when temperature rises.

Road and railroad buckling Figure 7.3

a. Excessive heat in the Midwest during July 2011 caused many roads to buckle.

Compressive forces cause buckling.

b. Railroad tracks expand when heated and can fail by buckling, a phenomenon called sun kink.

Another common application of thermal expansion occurs in bimetallic strips. Components of thermometers and circuit breakers, bimetallic strips are also used to open and close electrical switches in thermostats. They consist of two flat lengths of two different metals that are bonded together. The two metals have differing values of thermal expansion. For example, brass and steel are often used because brass expands more than steel when heated, causing the strip to bend. The strip bends toward the side with the lower thermal expansion—steel, in this case.

We can quantify thermal expansion in one, two, or three dimensions. Thermal expansion ΔL of a length of a rod depends on the temperature increase, ΔT, the original length L at the starting temperature, and the coefficient of thermal expansion α

$$\Delta L = \alpha L \Delta T$$

The changes in surface area ΔA or volume ΔV follow similar relationship:

$$\Delta A/A = 2\alpha(\Delta T)$$

and

$$\Delta V/V = 3\alpha(\Delta T)$$

Note that the area coefficient of expansion is equal to twice its linear value and the volumetric coefficient of expansion is equal to three times its linear value for most solids.

There is one important exception to the thermal expansion of matter with increasing temperature. Water expands slightly as it is cooled from 4°C to 0°C. **Figure 7.4** shows a graph of water's volume at different temperatures just above 0°C.

Volume of 1 g of water at temperatures around 4°C Figure 7.4
As water cools from 4°C to 0°C, it does not contract but expands, becoming less dense, and remains at the surface as it cools to 0°C.

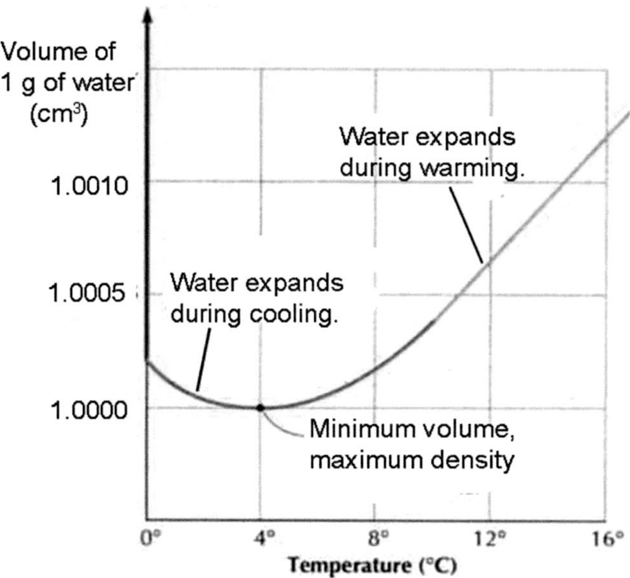

Water's expansion during cooling below 4°C means that water freezes from the top of a pond down, and not from the bottom up, as we might expect. As the temperature drops below 4°C, the water molecules take on the crystalline structure of ice, with more volume between molecules. Ice is less dense than liquid water and floats at the surface, allowing ice to form at the top of a pond.

If water continued to shrink as it cooled, the cool water would continue to sink as the temperature approached the freezing point, and it would freeze from the bottom up. This would be disastrous for all the aquatic life in the pond. The fish would be frozen in or pushed to the freezing surface.

The increase in volume of ice is responsible for bursting water pipes in homes during the winter. If we allow water to freeze inside a pipe, the pipe will rupture. Also, freezing water in the cracks and potholes in roads during the winter widens them and makes them larger, as you may notice while driving each spring.

CONCEPT CHECK

1. **How** does heat differ from thermal energy?
2. **Why** does sand heat up faster than water?
3. **Why** would you leave extra space in a water jug before freezing it?

7.2 Heat Transfer

LEARNING OBJECTIVES

1. **Describe** heat transfer by conduction.
2. **Explain** heat transfer by convection.
3. **Describe** heat transfer by radiation.

If you ever ran barefoot across the hot sand on a beach, you probably raced for the water as fast as you could to reduce the burning sensation on the soles of your feet. Sand transfers heat better than water, so it feels hot and water does not. We will describe here how thermal energy flows between objects and how it affects our perception of what is hot or cold

Objects transfer thermal energy to and from the environment in three different ways: **conduction**, **convection**, and **radiation**. Conduction is the principal heat loss mechanism within a substance or between substances that are touching. Convection is the principal mechanism when a fluid circulation pattern exists. Radiation dominates when an object is much hotter than its surroundings or when heat loss via conduction and convection is suppressed, as in the case of an object surrounded by a vacuum. See **Figure 7.5**.

Heat transfer by convection, conduction, and radiation Figure 7.5
Our hands above a fire are warmed by rising air via convection. Hands held to the side of fire are warmed by radiation. A metal bar held in the fire warms up by conduction of atoms and electrons in the metal.

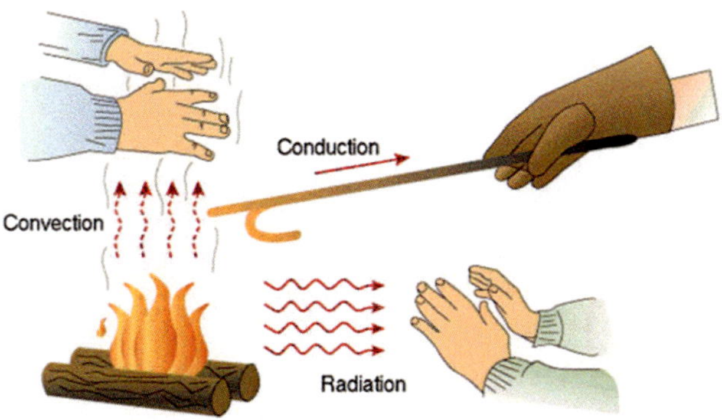

conduction Transfer of heat energy by collisions of atoms and electrons within a solid substance.

convection Transfer of heat energy by a moving liquid or gas.

radiation Transfer of heat energy by electromagnetic waves.

A simple cup of hot coffee sitting on a kitchen table cools down with the help of all three of these heat transfer processes. Let us examine each process in more detail.

Conduction

If you hold one end of a metal spoon and put the other end in boiling water, it does not take long before you feel warmth on your end. Cooks like to use wooden spoons rather than metallic ones because metals heat up quickly and burn upon contact with skin. Wooden objects, on the other hand, seem cooler because wood does not transfer heat well. When we touch a metal spoon and a wooden spoon, both placed in boiling water, the metal spoon feels hotter even though both spoons are at the same temperature. Higher thermal conduction is responsible for our perception of a hotter object.

On a microscopic level, as atoms in solids vibrate about their equilibrium positions, they push on each other. When we heat up one end of a metal rod, the atoms on the hot end vibrate faster and with increased amplitude. They bump harder into their neighbors and pass on part of their kinetic energy to them. The neighboring atoms then vibrate faster and with greater amplitude as well. This disturbance travels like falling dominos along the rod until it reaches the other end.

Some atoms have free outer electrons that can move easily within matter. They transfer energy by collisions with atoms and other electrons even more effectively than jiggling atoms can. Metals have plenty of these free electrons available and are excellent thermal conductors.

Poor thermal conductors are called *insulators*. Plastics, rubber, wood, and water are poor thermal conductors because they have few free electrons. For instance, the temperature difference between water or ice on the surface of a pond and the water at the bottom is due to water's poor thermal conductivity. **Figure 7.6A** shows an experiment that demonstrates water's low thermal conductivity. Poor thermal conductivity explains why the entire volume of water in a pond does not freeze, even when given enough time (**Figure 7.6B).**

Poor thermal conductivity of ice and water Figure 7.6

A. If you add boiling water to one end of a tube filled with ice, heat travels very slowly down the tube, and the remaining ice melts very slowly.

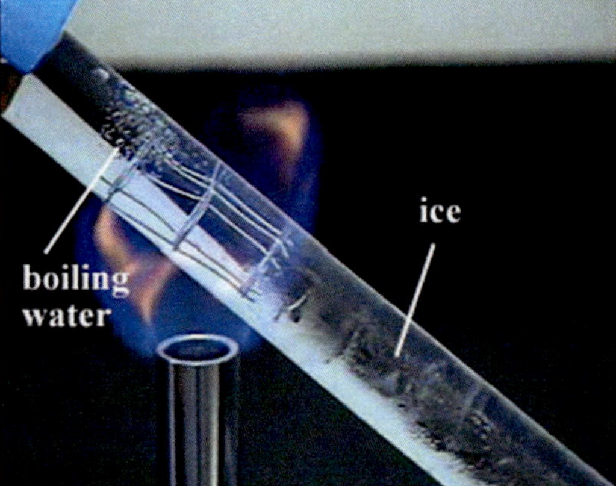

B. Poor thermal conduction of both ice and water is responsible for the pond not freezing in its full volume.

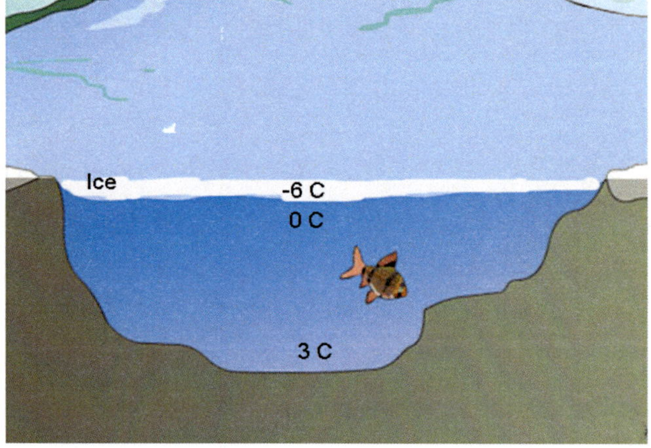

What determines how fast heat travels across an object? The rate of heat transfer by conduction is proportional to the temperature difference across the conducting object and the area of thermal contact. A larger area of thermal contact, as well as a shorter distance between the hot and cold regions, yields a greater transfer of heat (**Figure 7.7**).

Heat transfer by conduction through a window pane Figure 7.7
A larger surface area of the window pane, a bigger temperature difference across the pane, and thinner glass all increase heat flow from the warmer side to the colder side of the pane.

$$Q/t = k\,A\,(T_{out} - T_{in})/L$$

Q/t = Amount of heat Q flowing across the window pane in time t

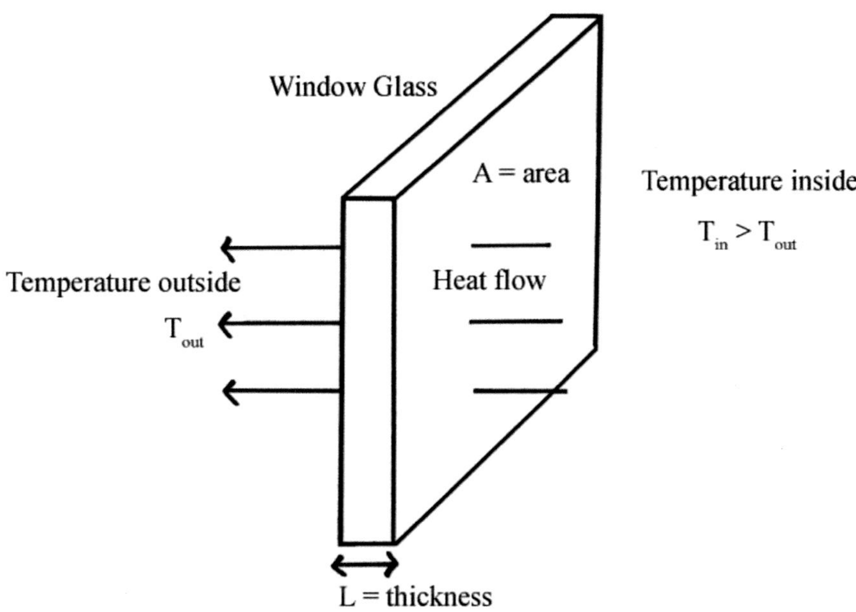

k = thermal conductivity constant = 0.8 W/mC for glass

The rate of heat flow varies across different materials. The property of a specific material that determines how well it conducts heat is called its *thermal conductivity*. **Table 7.2** lists the thermal conductivities of some select materials. The larger the thermal conductivity number, the more easily heat flows across the material.

Thermal conductivity k, in W/(m°C) Table 7.2

Metals	
Silver	406
Copper	386
Aluminum	210
Iron	73
Steel	46
Lead	35
Other solids	
Diamond	1000
Ice	2.2
Concrete	1.1

Fat	0.2
Skin	0.21
Window glass	0.8
Wood	0.08
Fiberglass	0.04
Polyurethane foam	0.024
Liquids and gases	
Water	0.58
Stationary air	0.026
Argon	0.016

The thermal conductivity of gases is smaller than for liquids and solids because the molecules in a gas are farther apart and collide less often. Stationary gas, such as still air, is an excellent insulator. Mountain climbers dig snow caves to escape severe weather and save themselves because still air in the cave and air trapped in snow reduce their body heat loss. The different layers of our clothing trap air that provides additional insulation against the cold. Birds ruffle their feathers and dogs and other fur-bearing animals bristle to capture a bigger layer of air between the skin and their feathers or fur.

Contemporary windows often have double panes and hold air, argon, or krypton gas between the panes to reduce heat loss. However, builders do not use air to insulate homes because once air starts moving, it loses its thermal insulating property and becomes an excellent heat conductor. Porous materials with pockets of trapped air, such as fiberglass or polyurethane foam, are great thermal insulators and are used instead.
Both skin and fat underneath the skin have low thermal conductivity, about a third that of water. Therefore bodies of humans and animals alike are well protected against sudden heat loss or heat gain, as **What a Physicist Sees** illustrates.

What a Physicist Sees
Firewalking

During firewalking, the temperature of the coals can be as high as 800°C (almost 1500°F). Yet a physicist realizes that the transfer of heat from the coals to the soles of the feet can be small if done properly. The firewalker can touch the coals briefly without any danger of burning his feet because coals—in particular, wood coals—have low thermal conductivity.

A shallow bed of coals cools off quickly, and moisture on the feet absorbs some of the heat that would otherwise go into the soles. Thick skin and walking quickly further reduce the amount of heat transferred during each step.

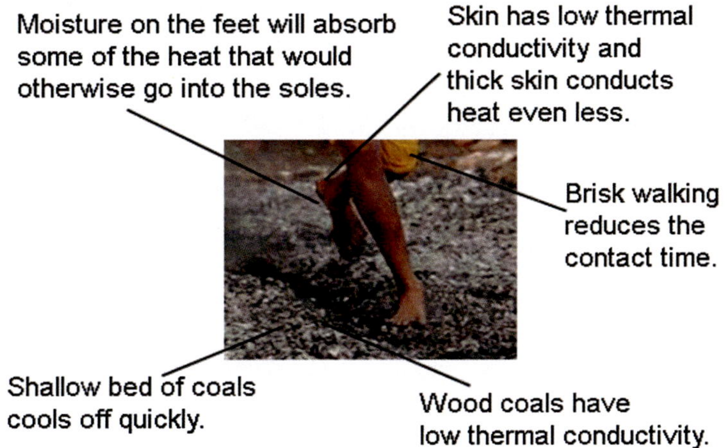

Convection

When we heat a volume of air, it expands and rises, carrying its thermal energy along. We call this transfer of heat **convection**. Chances are your home is heated by convection. Movement of warm air transfers thermal energy within the room.

Convection currents occur in the atmosphere and in the oceans. Gulf Stream in the Atlantic Ocean along the East Coast of the United States transfers heat from south to north and is an example of convection current.

A cool sea breeze on a beach is another example. Sand on the beach has a small specific heat, about six times smaller than that of water. It takes much less heat to warm 1 gram of beach sand by one degree Celsius than to heat one gram of ocean water, so sand warms up much faster than the water. The hot beach sand heats up the air in contact with it, and the warm air rises. Cooler air above the water moves in to replace the rising warm air over the beach. The rising air cools and then drops, replacing the cool air that moved in toward the beach. The result is a cooling on-shore breeze. At night, the cycle reverses and we get an off-shore breeze. See **Figure 7.8**.

Convection currents provide a mechanism for the transfer of heat within a body of water inside a pot on a hot stove. Water at the bottom of the pot heats, expands, and rises. At the same time, cooler and denser water on the surface sinks and replaces it. A convection loop develops and enables a faster transfer of heat from the bottom of the pot to the rest of the water. Without it, we would have to rely on water's low thermal conduction and heating would be much slower.

Convection also helps speed up the flow of heat from our bodies. Air coming in contact with the skin warms, rises, and carries away some of our body's thermal energy with it. This heat flow can be greatly increased in windy conditions because wind quickly replaces the warm air in contact with our skin and replaces it with cooler air. This is why people cover even their faces in cold, windy conditions to reduce their body heat loss.

Process Diagram
Land-sea air convection Figure 7.8

A. Heat transfer by convection during the day. Land warms up faster than the sea. Air in contact with the land warms and rises, to be replaced by a cool breeze from the sea. The result is an on-shore breeze.

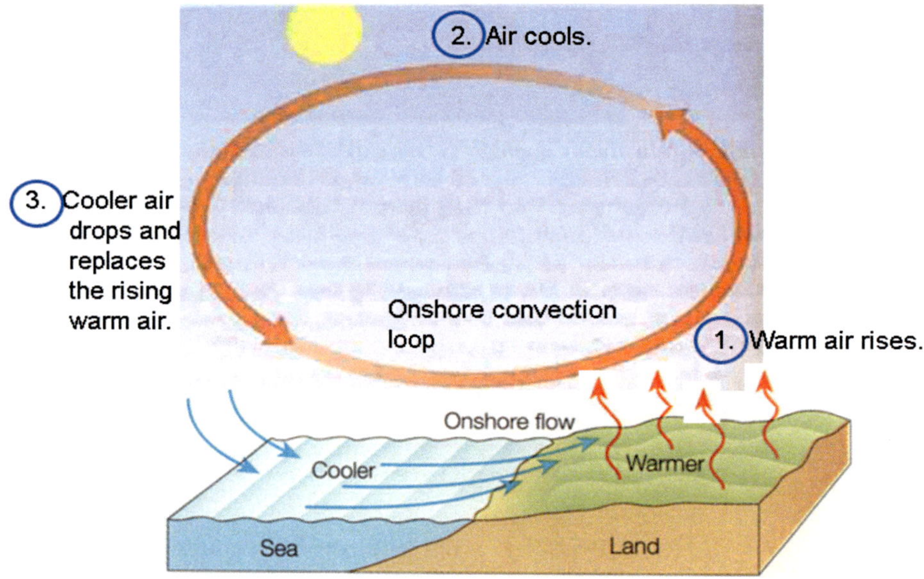

B. Heat transfer by convection at night. At nighttime, the heat flow reverses because the land cools faster than the sea. Air in contact with the sea warms and rises, to be replaced by a cool breeze from the shore. The result is an off-shore breeze.

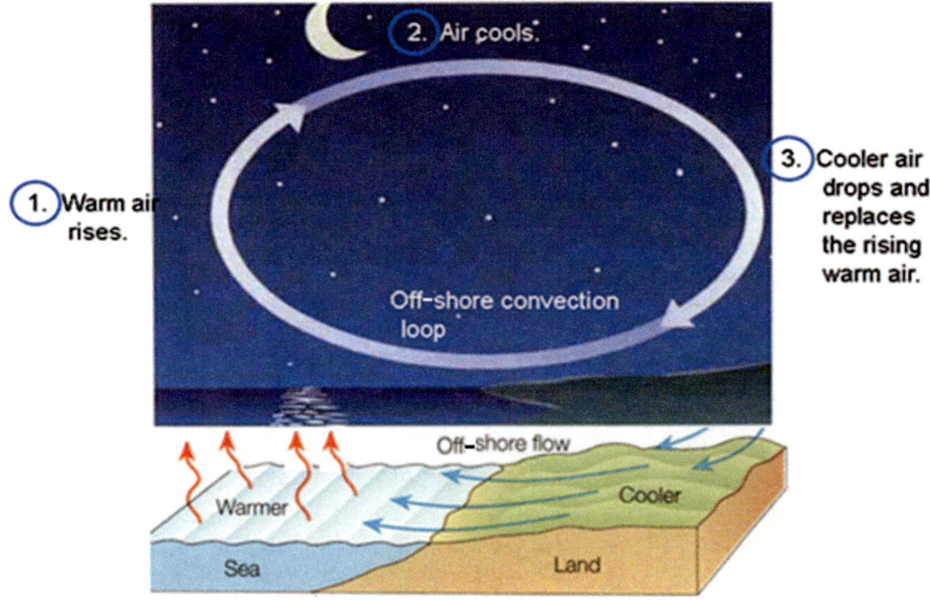

Radiation

When you lie on the beach and feel the sunlight warm your body, you experience heat transfer by radiation. **Radiation** of heat occurs through *electromagnetic radiation*, which can travel even through empty space. Thus, radiation is effective where heat transfer by conduction or convection fails. For instance, heat transfer by conduction and convection fails in the empty space between the upper atmosphere of Earth and the Sun. The Sun is just one source of radiation. All objects, hot or cold, radiate energy; however, most radiation is not visible because it falls outside the visible part of the electromagnetic spectrum.

Radiation is not a one-way process. If it were, all objects would eventually cool down to absolute zero. In addition to radiating energy, all objects receive radiated energy from their environment. The two opposite heat flow rates may or may not be equal. When an object radiates more energy than it receives from the environment, it cools down, and when it receives more energy than it radiates, it warms up. Objects in thermal contact, given sufficient time, reach the same temperature.

Not all objects absorb and emit radiant energy equally. A black object is a good absorber of radiant energy and reflects very little of the incoming radiation. Good absorbers of heat are also good emitters of heat; otherwise they would keep getting warmer and warmer. Each object at a constant temperature is in equilibrium, emitting as much energy as it absorbs. Black, rough materials emit radiation energy most easily, and polished metals or white objects radiate energy with difficulty, at rates that are up to 30 times smaller than for black objects. Black asphalt roads absorb more energy per square meter each second than a white snow field, just as a black car heats up more quickly than a white one. For the same reason, you want to wear white clothing when it is hot and dark clothing when it is cold. Also, a black roof absorbs more solar energy during summer and loses more heat during winter. A lighter color can keep your home cooler during the summer and warmer during the winter.

We can describe the radiation heat transfer rate Q/t by

$$Q/t = \sigma A \varepsilon T^4.$$

Here σ is the Stefan-Boltzmann constant, $\sigma = 5.67\ 10^{-8}$ W/m^2 K^4, A is the surface area of the object in square meters, ε is the emissivity, and T is the temperature in kelvin. Note that emissivity ε is a dimensionless number between 0 and 1. It is a ratio of the energy an object actually radiates to the energy the object would radiate if it were a perfect emitter.

We describe different types of radiation by their wavelengths (examined in detail in the next chapter). To get a simplified idea of wavelength, imagine producing different vibrations by shaking up and down the loose end of a rope attached to a wall. You will see a series of pulses traveling down the rope. The distance between any two successive pulses defines their wavelength. See **Figure 7.9**. Notice that if you move the rope lightly with little energy, its wavelength is longer, whereas if you pump it more vigorously with more energy, its wavelength is shorter.

Earth is constantly absorbing solar short-wave (higher energy) radiation and emitting long-wave (lower energy) radiation. In this balance, the Sun provides a nearly constant flow of short-wave radiation toward Earth, which is received at the top of the atmosphere. Part of this radiation is scattered away from Earth and heads back into space without being absorbed. Clouds and dust particles in the atmosphere contribute to this scattering. Land and ocean surfaces also reflect some short-wave radiation back to space. The short-wave energy from the Sun that is not scattered or reflected is absorbed by the atmosphere, the land, or the ocean. The absorbed solar energy raises the temperature of the atmosphere as well as the temperatures of the ocean and land surfaces.

Wavelength of radiation model Figure 7.9
Radiation is described by its wavelength, which can be modeled by the vibration of a rope attached to a wall.

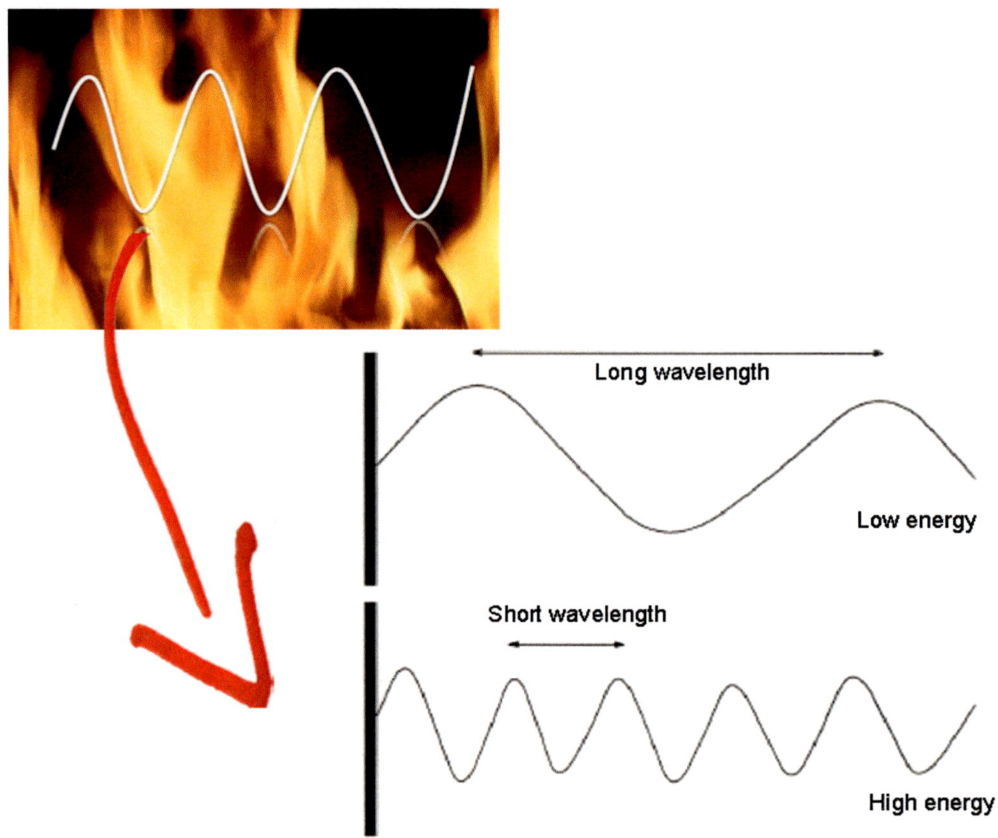

At the same time, the atmosphere, land, and ocean also emit energy in the form of long-wave radiation. This radiation ultimately leaves the planet, headed for outer space. The long-wave radiation outflow represents a loss of energy and hence tends to lower the temperature of the atmosphere, ocean, and land, thus cooling the planet. In the long run, these flows balance—the incoming energy absorbed equals the outgoing radiation emitted. Since the temperature of an object is determined by the overall (or net) amount of energy it absorbs and emits, Earth's overall temperature tends to remain constant as long as these two energy flows are in balance.

When directly overhead, the Sun radiates about 1400 J of energy onto each square meter of the upper atmosphere each second. In power units, this is 1400 W/m^2. This input energy is called the **solar constant**. This radiant energy is reduced by absorption in the atmosphere and by the non-perpendicular angles of the radiation as the Sun moves across the horizon. When we also take into account that we do not get any sunlight at night and less in winters than summers, the average amount of solar power on each square meter of surface in the United States is about 180 W/m^2 when averaged over a year.

solar constant The amount of solar radiation arriving perpendicularly to each square meter of Earth's upper atmosphere, equal to 1400 W/m^2.

Over the past few decades, the balance between the radiation that Earth receives from the Sun and emits back into outer space seems to be disturbed. We can use a greenhouse analogy to explain what has happened. In a greenhouse, glass (and its plastic equivalents such as Plexiglas) blocks the long-wave radiation that objects emit. The objects inside a greenhouse absorb the short-wave solar radiation and re-emit lower-energy, long-wave radiation. But this radiation cannot pass easily through the glass and the greenhouse warms up. This is the principle of the greenhouse.

We know that the Sun's radiation is absorbed by Earth's surface and is transformed into thermal energy. Earth re-radiates this energy in the form of longer-wave radiation. But if this radiation (also called terrestrial radiation) left Earth as easily as it came, then at night Earth's temperature would fall far below zero very rapidly. The reason this doesn't happen is that Earth's atmosphere contains gas molecules, called greenhouse gases, that absorb the heat and re-radiate it in all directions, some of it back to Earth. Greenhouse gases act like the glass panes of a greenhouse, causing Earth to remain warm enough for life to thrive on our planet.

Carbon dioxide (CO_2) is one of the greenhouse gases. It consists of one carbon atom with an oxygen atom bonded to each side. When these atoms are bonded tightly together, the carbon dioxide molecule can absorb long-wavelength radiation, and the molecule starts to vibrate. Eventually, the vibrating molecule emits the radiation again, and it is likely absorbed by yet another greenhouse gas molecule.

Water vapor (H_2O), methane, and nitrous oxide are a few of the other greenhouse gases that trap re-radiated terrestrial radiation. The greenhouse gases serve to hold heat in like the glass walls of a greenhouse. This absorption-emission-absorption cycle keeps Earth's heat near the surface, effectively insulating the surface from the cold of space. These gases are responsible for the fact that Earth enjoys temperatures suitable for our active biosphere. Without the greenhouse effect, life on Earth as we know it would not be possible because Earth's temperature would fluctuate wildly from extreme hot to extreme cold as day turns into night. In a sense, the greenhouse gases provide a blanket around Earth that moderates large temperature extremes during the night-day cycle.

While the normal greenhouse effect is beneficial, our concerns are related to the possible impacts of an *enhanced* greenhouse effect. Human activities, such as the burning of wood, coal, and oil, have increased the levels of greenhouse gases, leading scientists to think that we are experiencing *global warming* as a result. This is still a controversial topic. Curtailing human activities that produce greenhouse gases may reduce global warming, yet it would profoundly affect the global economy. Global warming will almost certainly remain a contentious issue in the future.

Many other examples of how we control or use heat transfer by radiation can be found in our daily lives. For instance, a thermos bottle has a double glass shell with a vacuum in between. The glass shell is coated with silvery reflective material and is protected by an outer casing, usually made of plastic or metal. The vacuum prevents any energy transfer through conduction and convection. The silver coating reflects much of the radiation, so radiation energy transfer is also minimized. Thus, the contents inside the vacuum flask can be kept at a more or less constant temperature for a long time.

Heat transfer in our homes

Many homes use a hot-water radiator for heating. It consists of a sealed, hollow metal pipe container that includes fins to increase the heating surface area. The water is heated by a boiler or furnace, using natural gas or heating oil. A circulating pump carries the water to all the radiators in the house. The hot water transfers its thermal energy to the radiator by conduction. As the pipes in the radiator heat up, they transmit their thermal energy to the room mostly by radiation and conduction. Once the water transfers its heat, the water cools, sinks to the bottom of the radiator, and is forced out of a pipe at the other end. Within a house, radiation is the primary mode of heat loss; conduction and convection are secondary and come into play only as matter interrupts or interferes with radiant heat transfer. In order to retard heat flow by conduction, walls and roofs are built with internal air spaces. Conduction and convection through these air spaces combined represent only 20%–35% of the heat that passes through them. In both winter and summer, 65%–80% of the heat that passes from a warm wall to a colder wall or through a ventilated attic does so by radiation.

Scientists can view objects or structures with a thermal scanning camera, which converts radiation emitted from any object into an array of colors on a color monitor. The image, called a thermograph, graphically maps the temperature into color, with red the hottest and blue the coldest. House energy efficiency inspectors use the cameras to diagnose heat losses. The thermograph technique has also been used in medicine. Unusually warm parts of a body may indicate cancerous growths, for instance. Night vision goggles work similarly; they convert infrared radiation into visible light. At night, they allow us to see human beings, who emit infrared radiation at higher temperatures than the radiation emitted by their environment.

CONCEPT CHECK

1. **What** does the rate of conduction between two objects depend on?
2. **How** does air moved by a fan help you cool off on a hot day?
3. **How** can you increase radiation heat transfer?

7.3 Changes of State

LEARNING OBJECTIVES

1. **Describe** how matter changes state.
2. **Explain** how our bodies remove excess heat.

Matter can exist in three different forms (also known as *phases* or *states*): solid, liquid, or gas. Energy is absorbed or released in transforming matter from one state to another. For instance, we must add energy to an ice cube to melt it, add more energy to warm up the resulting water, and still more energy to vaporize it. Molecules in the different states have different spatial arrangements. The added energy goes into loosening and breaking bonds between the molecules and increasing the molecules' kinetic energy. This energy is released if the substance changes state from gas to liquid or from liquid to solid.

Energy and Change of State

Take an ice cube and add heat to it. Eventually, all of it melts and turns into liquid. If we keep adding more energy, the water will boil, and eventually the water will turn into vapor. We can reverse the process by removing heat from the water vapor to condense it (**Figure 7.10**).

The amount of energy needed for a substance to change from one state to another is called *latent heat*. Whether the change is from a liquid to gaseous state (**evaporation**) or from a gaseous to liquid state (**condensation**), the latent heat is the same; the same holds for melting versus freezing. See **Figure 7.11**.

evaporation Change from a liquid state to a gaseous state.

condensation Change from a gaseous state to a liquid state.

Condensation on a cold glass Figure 7.10
When warm air comes in contact with a cold surface, water vapor in the air condenses, changing from a vapor to a liquid.

Change of state in water Figure 7.11
To change from one state into another, we either add or subtract energy called latent heat.

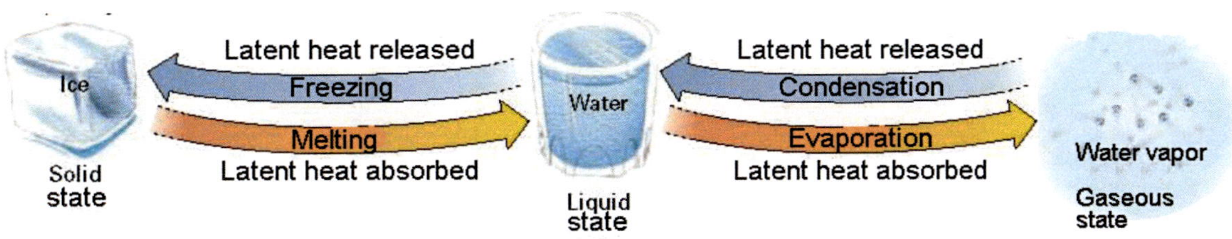

Let us examine the process of converting a cold ice cube into water vapor. Starting with the ice cube at a subfreezing temperature of –10°C, we heat it up. The cube warms and reaches 0°C, the melting point of ice. We keep providing heat, but the temperature temporarily stops increasing. Now all the added energy goes into breaking the bonds of the ice cube's crystalline structure. That is, the energy goes into changing the ice into a liquid state. The ice-water mixture remains at 0°C until all of the ice melts away. It takes about 334 joules of energy to melt 1 g of ice at 0°C into 1 g of water at the same temperature. We call this energy the **latent heat of melting** *L* for water.

latent heat of melting Amount of energy required to change a unit of mass of a substance from a solid to a liquid phase.

Once all the ice is converted to water at 0°C, then additional heating of the water increases the kinetic energy of the water molecules and the water becomes warmer. It takes about 4.2 joules of energy to increase the temperature of 1 g of water by 1°C. When the temperature of water reaches 100°C, its boiling point, the temperature stops increasing again, even though we keep adding energy. Now the energy goes into breaking the bonds between the water molecules. This energy is needed so the molecules can leave the surface of the water and enter a gas phase. More heat is required here than for melting. It takes 2,257 joules to change 1 g of water at 100°C into 1 g of vapor at the same temperature. This energy is called the **latent heat of vaporization**.

latent heat of vaporization Amount of heat required to change a unit of mass of a substance from a liquid to a gaseous phase.

The heat Q that must be supplied or removed to change the phase of a mass m of a substance is $Q = mL$.

Once all the boiling water converts into vapor, the vapor becomes hotter as we add more heat. The kinetic energy of the water vapor molecules increases and their temperature rises as well. See **Figure 7.12** on the next page.

Change of state of ice into water and then into vapor Figure 7.12

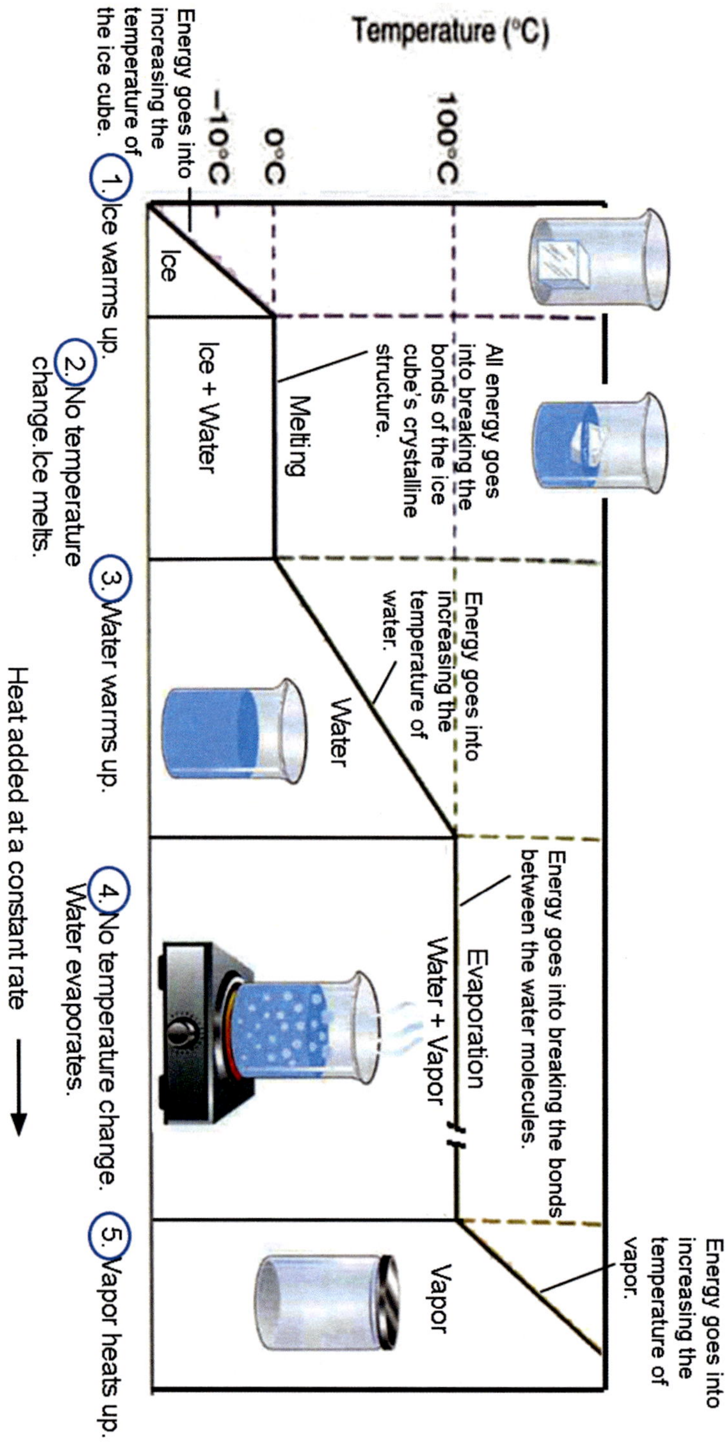

The latent heat of melting for ice is 334,000 J/kg, which is a relatively large number. This is why alpine snow, a collection of micro ice crystals, stays around in some mountain ranges way into the summer. If the latent heat of melting were much smaller, then warmer temperatures in the springtime would guarantee floods if several meters of snow suddenly melted.

You may wonder what role motion plays in freezing because streams often do not freeze in subfreezing temperatures. Motion of water by itself should not stop it from freezing. It is water above 0 °C that continuously feeds a stream and keeps its temperature above the freezing point (**Figure 7.13**).

Running water in subfreezing temperatures Figure 7.13
A continual replacement of water in a stream by water from a warmer constant-temperature source makes freezing of the stream difficult.

It is interesting to compare water to other liquids, such as liquid hydrogen, oxygen, nitrogen, and helium, all of which vaporize easily. At atmospheric pressure, helium vaporizes at 4.2 K, hydrogen at 20 K, oxygen at 90 K, and nitrogen at 77 K. The latent heat of vaporization of liquid nitrogen is 199,000 J/kg and of liquid helium is 21,000 J/kg. If you pour out liquid nitrogen or helium out of a storage dewar, they will vaporize almost instantly. Imagine if this happened to water.

The large amount of heat needed to vaporize water, 2,257,000 J/kg, has been critical in sustaining life. If water's latent heat of vaporization were not this high, water would exist mostly in its gaseous state, like oxygen and nitrogen.

There is a close link between boiling and vaporization. Boiling of water can be thought of as vaporization taking place throughout the volume of the liquid. It begins when the pressure of water vapor within the liquid water (vapor pressure) is equal to the atmospheric pressure. This occurs at 373 K (100ºC). If the pressure inside a bubble is equal to atmospheric pressure, the bubble rises and eventually escapes as vapor. See **Figure 7.14**.

When we heat water in a pot with a lid on, the pressure of the water vapor above the water surface continuously increases, preventing some more energetic water molecules from leaving the surface of the water and entering the vapor state. As a result, the boiling temperature increases. Only when the pressure reaches a threshold point where it can lift the lid on the top of the pot can the steam leave and the pressure stop increasing. The boiling temperature of water can rise to 120ºC in a pressure cooker, and food cooks faster.

Boiling of water Figure 7.14
During boiling, gas bubbles form within the liquid, then rise and escape the liquid.

Reduction of pressure has the opposite effect on the boiling temperature. When we boil water in the mountains, we need to cook our food longer because water boils at a lower temperature at higher altitudes, where the air is thinner. For instance, it boils at 94.15 °C in Denver, Colorado, which has an elevation of 1,610 meters (5280 feet). This is a 5.85 °C drop relative to the boiling temperature at sea level. If you reduce the pressure significantly, you can even get water boiling at room temperature. A beaker of water at room temperature (20 °C) in a bell jar will boil if you reduce the pressure inside the jar to about 2220 N/m^2 (17 mm Hg). See **Figure 7.15**.

Boiling of water at different pressures Figure 7.15
Water boils at a higher temperature when under higher pressure than normal and at a lower temperature when under lower pressure than normal.

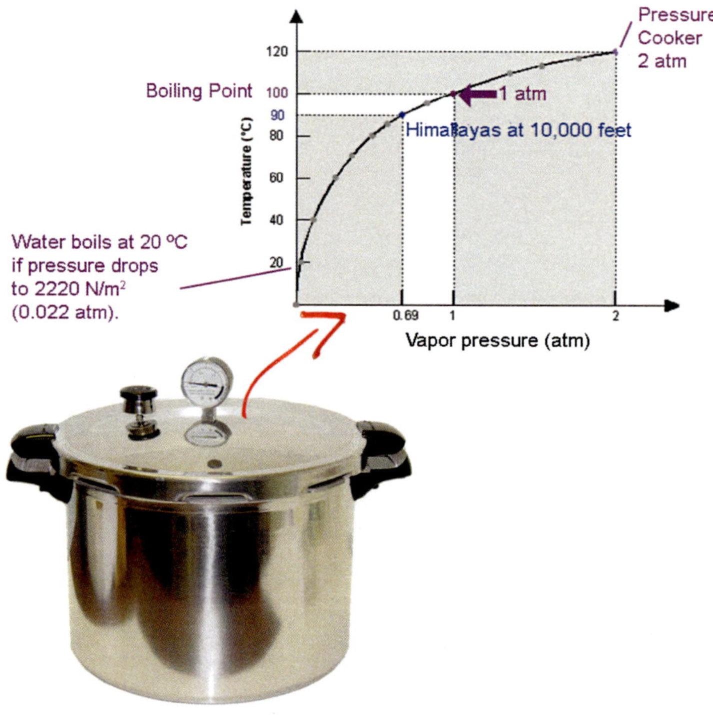

You do not have to have boiling to have evaporation. Different amounts of water evaporate at all temperatures between 0 and 100°C. Since energy leaves the evaporating body, this process is also called *evaporative cooling*. For instance, the evaporation of sweat during running reduces your body temperature.

Blowing on a cup of hot tea or coffee is another example. Water molecules are continuously leaving and entering the surface of a liquid. The escaping molecules remove energy from the liquid, while the arriving molecules add energy to the liquid. In equilibrium, the two rates are equal. If we decrease the density of the water vapor above the liquid by blowing on it, condensation will decrease and the equilibrium is shifted. More molecules leave the liquid, resulting in overall cooling of the liquid. That is why you blow on a hot cup of coffee to cool it (**Figure 7.16**). Evaporative cooling is the idea behind rubbing alcohol, which is applied as a cooling, soothing application for bedridden patients or athletes. See **Figure 7.17**.

Evaporative cooling of coffee Figure 7.16

By blowing on the coffee, you remove the water vapor above the coffee, which enhances further evaporation and thus cools the coffee.

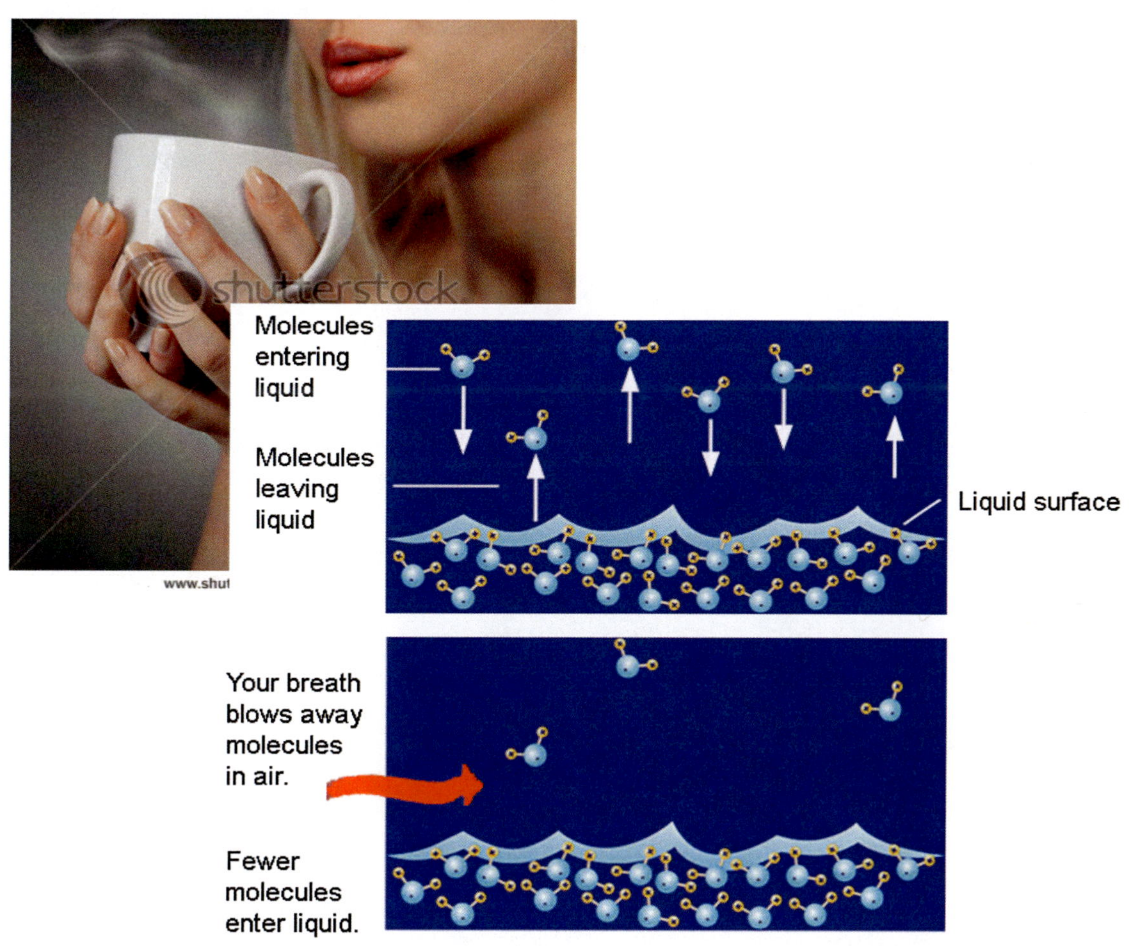

Cooling effect of rubbing alcohol Figure 7.17
Liquid alcohol chills faster than water because it loses molecules faster. Water has strong intermolecular forces, which make it bead up on a surface. Alcohol's intermolecular forces are weaker, which allows it to spread into a larger area. This larger area also enhances evaporation.

When exposed skin becomes wet, body heat evaporates the moisture, which chills the skin. At the same time, moist clothing draws heat away from our bodies because water is a much better thermal conductor than air. This is why a cold, damp day often feels colder than a cold, dry day at the same temperature. During heavy exertion or in hot weather, evaporation becomes a major way of losing body heat. The body is able to increase its evaporation by sweating. For instance, when we are running, the metabolic conversion of chemical energy into heat energy can be as high as 500 joules per second, or 500 W. We need to unload this increased internal energy quickly. Sweating profusely does just that job.

CONCEPT CHECK

1. **What** happens, at the molecular level, during melting?
2. **Why** does sweating cool our bodies?

7.4 Heat, Work, and Thermodynamics

LEARNING OBJECTIVES

1. **Explain** the first law of thermodynamics.
2. **Describe** an adiabatic process.

We're all used to the idea of boiling water to make tea or coffee, but wouldn't it seem strange if we had to boil water every time we wanted to do anything? What if we had to make steam to charge our cell phone, watch TV, or dry our hair? It may sound bizarre, yet it's not so far from the truth. Most of electrical power we consume in our homes probably comes from a power plant that generates electricity from boiling, hissing, and rapidly expanding pressurized steam.

Invented at the dawn of the Industrial Revolution in Britain, the steam engine generated tremendous power and liberated both humans and animals from the drudgery of muscle-powered labor. Steam engines and later steam turbines found a way of converting energy stored in coal into useful work, using heat. The interplay of heat and work evolved into a new scientific discipline, called **thermodynamics**.

Thermodynamics provides a foundation for understanding the conservation of energy It also tells us that heat flows spontaneously from hot to cold and not from cold to hot. Here we will study the conversion of thermal energy into work and examine what is possible and what is not during the energy conversion processes. As we will see, the useful mechanical energy in many processes eventually transforms into thermal energy that becomes difficult to re-use.

thermodynamics The study of heat and its transformations to different forms of energy.

Heat played an important role in formulating the idea of conservation of energy. The conversion of energy between the kinetic and potential energies of a falling object was easily visualized and understood. However, the conversion of energy between these two forms and heat was not so obvious because heat involves the interactions of randomly moving and colliding molecules, atoms, and electrons, not visible to the human eye. During the 19th century, scientists realized that heat was a form of energy able to change into other forms. This idea brought about the acceptance of the conservation of energy as a general physics principle.

Mechanical equivalent of heat

German physician and physicist Julius Robert Meyer (1814-1878) is credited with first proposing an equivalence of all forms of energy, including heat, and the conservation of total energy. However, Englishman James Prescott Joule (1818_1889) confirmed the energy conservation principle in his experiments by measuring heat transfer. In his now famous experiment, Joule designed a clever way of doing mechanical work on a system consisting of a paddle wheel inserted in an insulated beaker filled with water. The amount of work had a consistent effect in raising the water's temperature.

In Joule's experiment, a falling weight suspended over a pulley turned a paddle wheel, which did work on the water in an insulated beaker. Here the gravitational potential energy of the weight is converted into the kinetic energy of the water molecules, increasing the internal energy of the water (Note that we use the term "internal energy" here rather than "thermal energy," used above, because physicists traditionally choose internal energy over thermal energy when describing thermodynamics.) A thermometer inserted in the water recorded the resulting change of temperature. Joule found that 4.186 J of work was required to raise the temperature of 1 gram of water by 1°C. See **Figure 7.18**. (At that time, the unit of joule [named in Joule's honor] did not exist, and he expressed his results in what are now antiquated units.)

Mechanical equivalent of heat Figure 7.18

Potential energy of a falling weight suspended over a pulley is converted into heat.

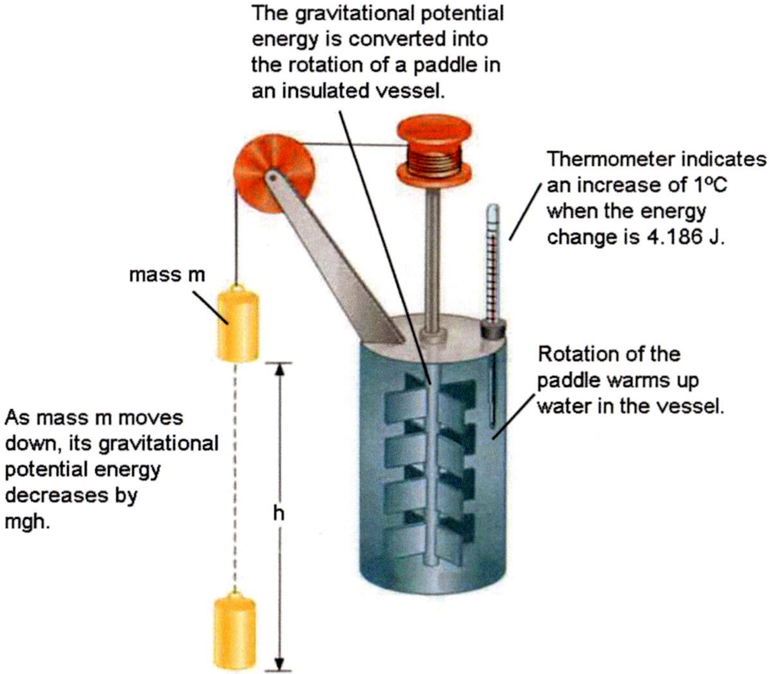

It was already known that 1 calorie is needed to raise 1 gram of water by 1°C. Therefore, this experiment showed the conversion between calories, which are used to describe thermal energy, and joules, which are units of energy used to describe work in general. James Prescott Joule designed several other experiments to measure the conversion of mechanical work to thermal energy, but the results were always the same: 4.186 J of work produced the same temperature increase as 1 calorie of heat. He showed that both heat and work represent transfers of energy into or out of a system, such as his beaker and paddle wheel apparatus. It was another step toward defining what heat is.

First law of thermodynamics

William Thomson (1824–1907), also known as Lord Kelvin, studied Joule's results and generalized a relationship between heat and work known as the **first law of thermodynamics**. This law is a statement of conservation of energy that includes heat. Thomson said that there are two ways to increase the internal energy of a system: heat it or do work on it.

The increase in internal energy of a system is equal to the heat added plus the work done on the system.

$$\Delta U = Q + W$$

Q is positive when the system gains heat and negative when it loses heat.

first law of thermodynamics The increase in internal energy of a system is equal to the heat added plus the work done on the system.

By "system," we mean a well-defined group of atoms, molecules, particles, or objects. By "work done," we mean a physical transfer of energy directly to atoms and molecules. The first law recognizes two ways to transfer energy: the mechanical way we know as work, and the thermal way we know as heat. For instance, the compression of the air-gasoline mixture in a car's cylinder increases its internal energy. Mechanically stirring water or heating it increases water's internal energy as well. See **Figure 7.19**.

Increasing internal energy of water Figure 7.19
You can increase the internal energy of water in two ways: one is to heat it and the other is to work on it by stirring it.

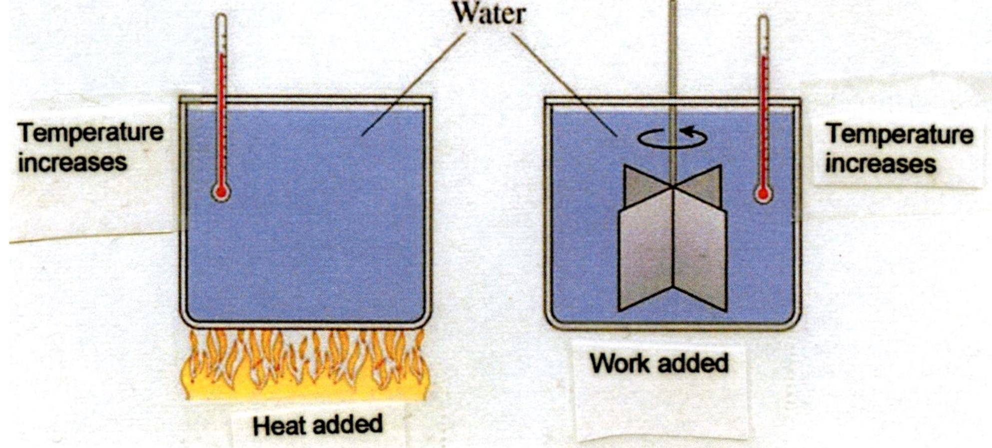

We could also restate the first law as *the decrease in internal energy of a system is equal to the heat removed plus the work done by the system*, as **Example 7.2** shows.

Example 7.2
Work and internal energy of a jogger

a) While jogging, you do 400,000 J of work and give off 320,000 J of heat. What is the change in your internal energy?
b) You stop jogging and cool down by walking, giving off 100,000 J of heat while your internal energy decreases by 250,000 J. How much work have you done while walking?

Solution:

a) Recall the first law of thermodynamics: the decrease in internal energy of a system is equal to the heat removed plus the work done by the system. You give off 320,000 J of heat and your body does 400,000 J of work. Therefore, your internal energy decreases by
320,000 J + 400,000 J = 720,000 J.
b) From the first law, we know that

Decrease in internal energy = work done by your body + heat removed or
250,000 J = work done by your body + 100,000 J

Therefore, you did 250,000 J – 100,000 J = 150,000 J of work while walking.

The first law states that the total energy of a system must include thermal energy, and this energy cannot be created or destroyed; it simply converts from one form to another. Prior to 1850, the concept of energy was restricted mainly to mechanics. The work of Joule and Lord Kelvin established that heat flow was a transfer of energy and had to be included in any energy accounting. This concept brought the energy conservation principle to the center stage of thermodynamics.

In many situations, compressing or expanding a gas may involve no heat flow in or out of a system. Such a process is said to be **adiabatic**. Adiabatic conditions occur when we insulate a system from its surroundings or perform some process very quickly, so that heat has little time to enter or leave. For instance, if we quickly compress a piston in a gas cylinder, there is little time to add any heat during the compression. Pumping air into a bicycle tire could also be considered adiabatic.

adiabatic process A process that occurs within a thermally insulating system. Fast expansion or compression of gas where no, or little, heat enters or leaves a system is modeled as adiabatic.

If no heat is added to or subtracted from a system, then the first law of thermodynamics tells us that *any work done on the system* goes directly into increasing the system's internal energy. But what does increasing the internal energy mean? You may have observed that after pumping air into your bicycle tire, the temperature at the base of the bicycle pump increased immediately. You increased the internal energy of the molecules within the gas, and in an ideal gas, this energy is all kinetic and is proportional to temperature. Therefore, by increasing the internal energy you also increase the temperature of the gas and the pump gets hotter. See **Figure 7.20A**.

Adiabatic heating and cooling Figure 7.20

A. When you pump air into a bicycle tire, you do work on the air to compress it into the tire, increasing the air's internal energy. The increase in internal energy of the air causes an increase in temperature at the base of a bicycle pump.

B. Air coming out of a person's pursed lips does work on the surrounding air and its internal energy decreases. The air expands past the lips and cools.

The opposite process also works. *If work is done by the system*, its internal energy decreases. For instance, if we allow the gas to expand and do work on the cylinder, the internal energy of the gas decreases and its temperature drops—a useful process if you want to cool something. See (**Figure 7.20B**). As a practical application, refrigerators use pressurized gas as a coolant. The gas is allowed to expand, which in turn lowers its temperature. This cool gas circulates through coils in the back of the refrigerator and extracts heat from inside the refrigerator. We will return to this idea later in the chapter.

Just as in the case of a bicycle pump (**Figure 7.20A**), the first law of thermodynamics explains what occurs inside the cylinder of a diesel engine. In the diesel engine, air is compressed adiabatically with a compression ratio typically between 15 and 20. This rapid compression of air increases the temperature of the gas enough to ignite the fuel mixture that is inserted in the cylinder. **What a Physicist Sees** shows another adiabatic process that may be familiar to many of you.

What a Physicist Sees
Adiabatic cooling

When a cold carbonated drink is opened, the expansion of the gas above the liquid is so rapid that the only possible energy transfer is from the internal energy of the gas. The gas loses internal energy and becomes colder, causing water vapor in the expanding gas to condense into water drops. You can see these drops as a slight fog at the opening of the bottle.

Critical Thinking...
How could you enhance this fogging effect?

CONCEPT CHECK

1. **How** can you change an object's internal energy?
2. **What** is the required condition to have an adiabatic process?

7.5 Second Law of Thermodynamics and Heat Engines

LEARNING OBJECTIVES

1. **Explain** the second law of thermodynamics.
2. **Apply** the second law of thermodynamics to heat engines.

We have all seen ice cream melting on a hot day. But why do we never see the reverse process? Why do we not observe the melted ice cream gradually refreezing? The first law of thermodynamics does not say anything about the direction of heat flow. As long as the energy conservation principle is satisfied—that is, the hotter object gains the same amount of energy that the cold object loses—the first law does not rule out the flow of heat from cold to hot. However, we know that such a flow does not occur spontaneously. Another principle must be operating to explain the flow of heat, putting a constraint on the first law.

Second law of thermodynamics

About fifty years ago, Sir Charles Snow, an influential British scientist and novelist, startled a gathering of nonscientists by saying that their lack of knowledge of the second law of thermodynamics was equivalent to scientists not ever reading a work of Shakespeare. Unfortunately, despite its importance, most nonscientists are not aware of the second law, even though it impacts us all every day.

The **second law of thermodynamics** simply puts a restriction on the first law by stating that processes that conserve energy only proceed in a certain direction. In effect, it identifies the direction of energy transformation in natural processes.

The second law of thermodynamics states that *It is impossible for heat to flow spontaneously from a colder body to a hotter body.* See **Figure 7.21.**

second law of thermodynamics The following three statements are equivalent: 1) Thermal energy never spontaneously flows from a cold region to a hot region. 2) No heat engine can be completely efficient in converting heat to work. 3) All systems tend to be more disordered over time.

Note that the second law does not prevent the flow of heat from a colder object to a hotter object. It just says that such a flow of heat does not happen on its own. We can extract heat from a colder body and place it in a hotter body by doing work. We will learn about this process when we examine the operation of refrigerators.

There are other, equivalent versions of the second law of thermodynamics. Let us see what the second law says about the conversion of heat to work. We already know from the first law that we can heat a cup of water by stirring it. In principle, we can convert all of this work into heat by insulating the cup. But how about converting heat into work? Can we do this? And if so, can we do it completely without any heat loss into the surroundings?

Direction of heat flow Figure 7.21

The second law of thermodynamics tells us what is possible and what is not: Heat flows spontaneously from a hotter object to a colder object. It does not flow spontaneously from a colder object to a hotter object.

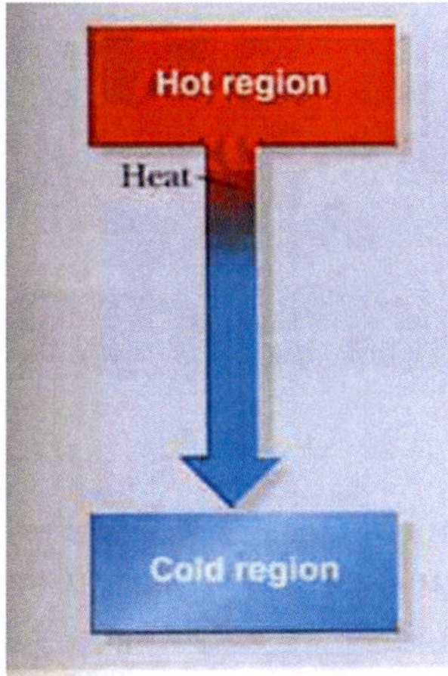

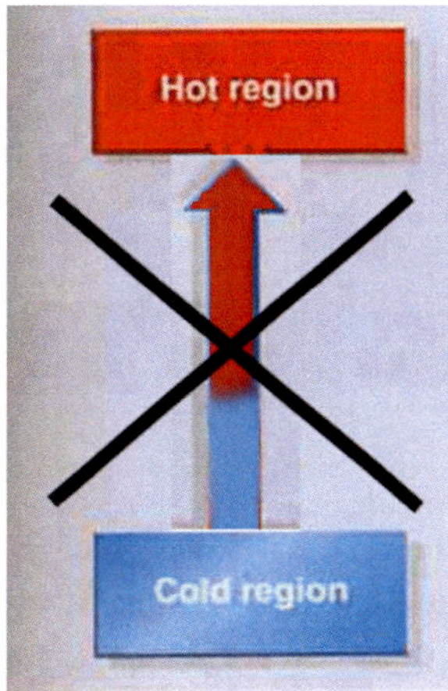

Heat engines

The first record of a primitive **heat engine** is from an ancient Greek mathematician and engineer known as Hero of Alexandria (10–70 AD) in Egypt. He converted a hot jet of steam coming out of a sphere's nozzle into rotation of the sphere. See **Figure 7.22**.

heat engine A device that takes in heat and converts part of it into useful work.

Hero's primitive heat engine Figure 7.22

Steam exhaust out of the nozzles causes rotation of the sphere. The motion of the nozzles works on the same principle as the propulsion of rockets.

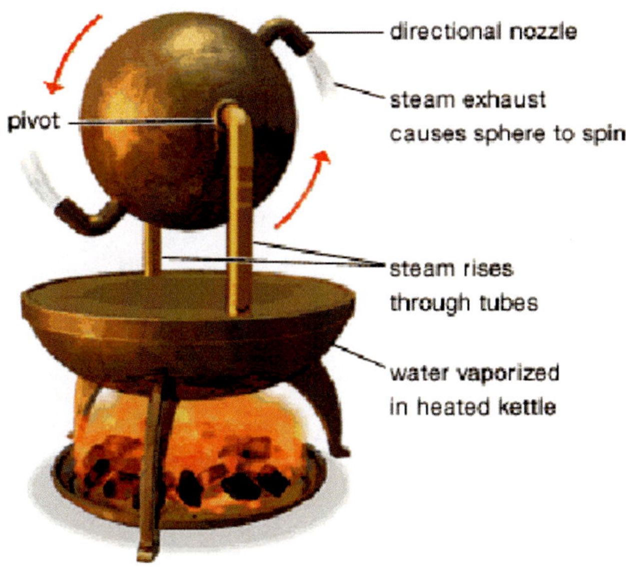

While Hero showed that converting heat into work is possible, converting heat into useful work required ingenuity. Many people in Britain worked on the design of heat engines during the 1700's and even earlier. These efforts culminated with James Watt (1736-1819) patenting his steam heat engine in 1769. It was the dawn of the British Industrial Revolution.

Steam locomotives were an impressive marvel of heat-engine engineering. The first prototypes were designed in England and the United States during the 1780s and 1790s. A locomotive is built around a steam engine and is based on a simple idea: you can burn fuel (coal) to release the energy stored in it. As the coal burns in a furnace, it releases heat, which boils water and generates high-pressure steam. The steam feeds through a pipe into a cylinder with a tight-fitting piston, which moves outward as the steam flows in—a bit like a bicycle pump working in reverse. The steam expands into the cylinder and gives up its energy to the piston. Movement of the piston pushes the locomotive's wheels around and the process repeats. Note that the steam is not a source of energy: it's an energy-transporting fluid that helps to convert the energy stored in coal into mechanical energy that propels the train. See **Figure 7.23**.

Steam-engine locomotives Figure 7.23

A. High-pressure steam expands into the cylinder and pushes on it.

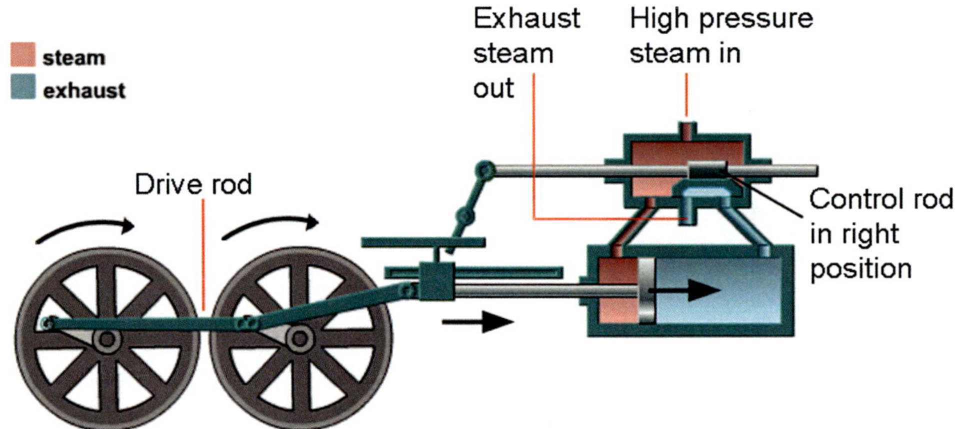

B. The control rod moves to the left, switching the steam input into the back of the cylinder, and the process repeats.

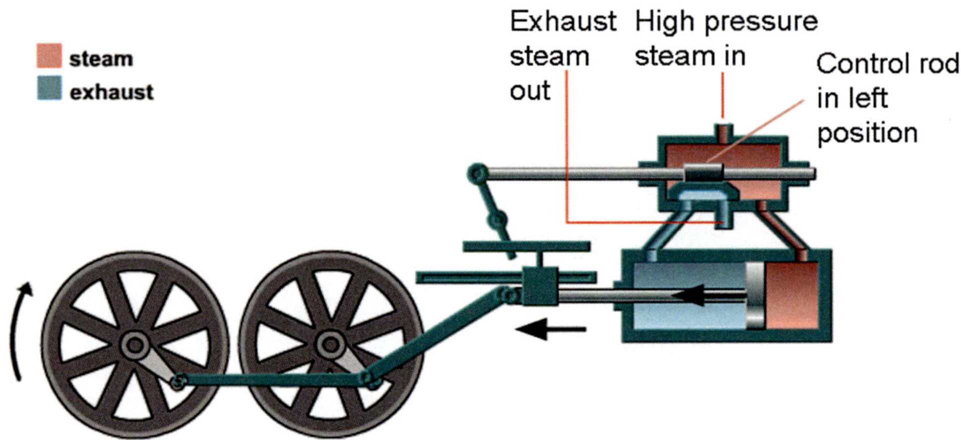

The movement of the piston is linked into a drive rod whose motion is converted into the motion of the wheels.

C. This is an ex-British Railways Standard 4MT locomotive (built at Brighton in 1955) working on the heritage Swanage Railway, England in 2008.

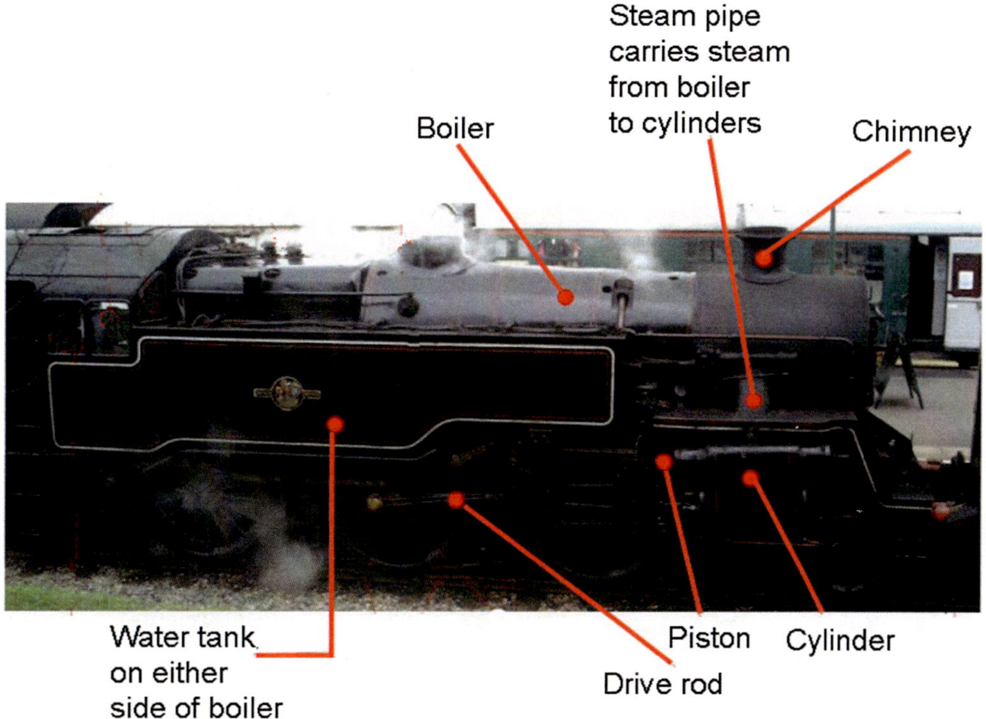

Steam engines powered the world throughout the Industrial Revolution, from the 18th century right up to the middle of the 20th century. But they were huge, cumbersome, and relatively inefficient. A simple, steam-driven piston and cylinder delivers energy to the machine it powers only 50% of the time (during the power stroke, when the steam is actually pushing it); the rest of the time, it's being pushed back into the cylinder to be ready for the next power stroke. Another problem is that pistons and cylinders produce back-and-forth, push-pull, reciprocating motion, when most of the time we prefer rotary motion—turning a wheel, say, without the pistons, cylinders, cranks, and all the rest. The steam turbine, an energy-conversion device perfected by British engineer Sir Charles Parsons in the 1880s, addressed that problem. It would directly power a wheel with the force of the expanding steam.

A turbine is a spinning wheel, as in a windmill or a fan that gets its energy from gas or liquid moving past it. Steam turbines use high-pressure steam directed over tilted blades, causing the blades to spin and to turn electricity generators at incredibly high speeds (**Figure 7.24**). A steam turbine is more compact than a steam engine: spinning blades allow steam to expand and drive a machine in a much smaller space than a piston-cylinder-crank arrangement would need. That's one reason why steam turbines were quickly adopted for powering ships, where space was very limited.

Process Diagram
Operation of a steam turbine Figure 7.24
High-powered steam causes the turbine to rotate. This rotational energy is converted into electricity.

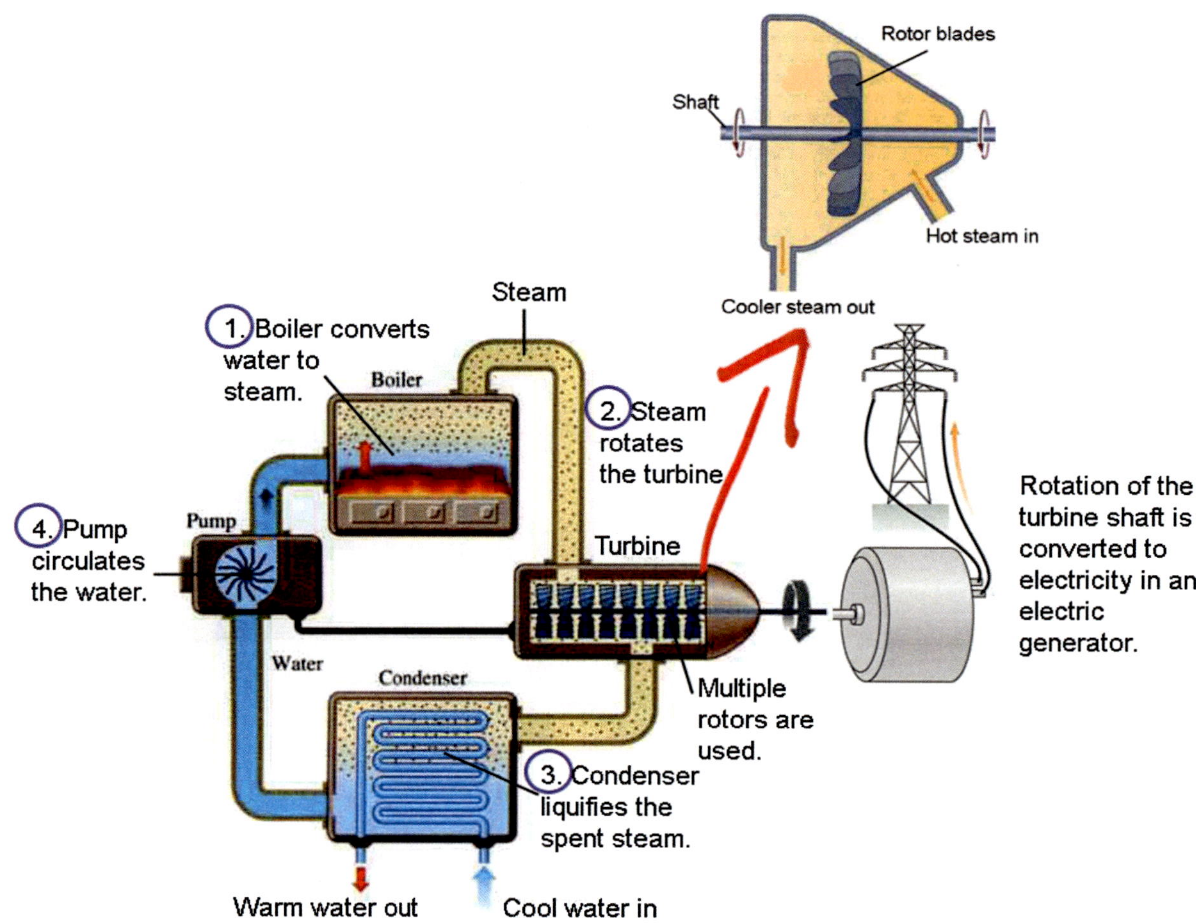

Although heat engines are designed to convert thermal energy into work, in practice they are capable of converting only *some* of the energy into work. The unused portion of the energy is discarded (or vented) as waste heat. Not all of the randomly moving gas molecules in the cylinder can participate in doing work. Only those moving in the right direction—against the piston—can do work. The rest bounce off the cylinder's walls and do not work on the piston. To even think of obtaining a perfect thermodynamic efficiency, we would need to line up the motion of all gas molecules so they always move straight against the piston. This is impossible because thermal energy is random energy.

We use the concept of heat reservoirs to describe the general operation of a heat engine. The term "hot reservoir" refers to any large source of thermal energy, such as the furnace of a steam engine or the hot gas driving a steam turbine. A "cold reservoir", such as the outside air or the ocean, accepts the waste energy of a heat engine after it has done its work. We consider both hot and cold reservoirs to be so large that their temperatures do not change during the operation of a heat engine.

Since the flow of thermal energy from a hot reservoir to a cold reservoir drives heat engines, we must have a difference in temperatures between two regions before we can have a heat engine. Heat flows from a high-temperature reservoir to a low-temperature reservoir, and along the way we harness some of the high-temperature heat into work, while the rest of the input heat goes to waste in a low-temperature reservoir. See **Figure 7.25**.

Energy distribution of a heat engine Figure 7.25
In a heat engine, part of the input heat coming from the hot reservoir is converted to work. The rest of the heat goes to waste in the cold reservoir.

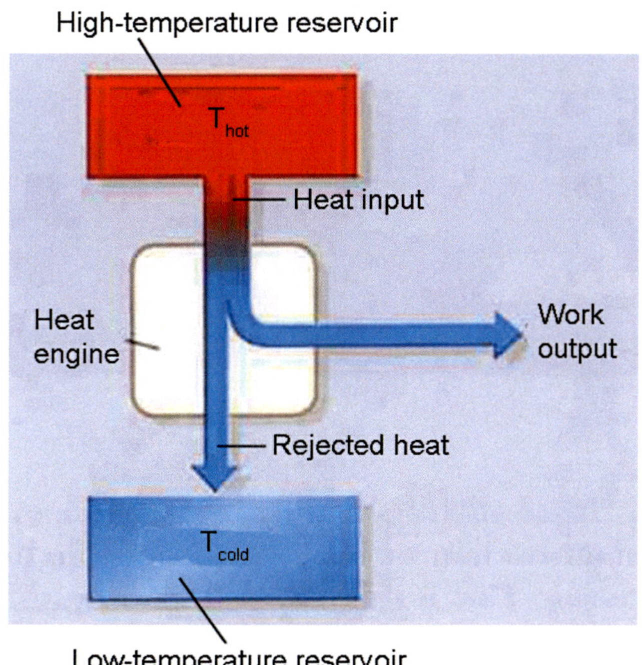

We can summarize the energy distribution of a heat engine as

Heat input = work output + waste heat output or

$Q_{in} = W + Q_{out}$

The waste output here is never zero. That, in essence, is the second law of thermodynamics as applied to heat engines:

It is impossible to have a heat engine that converts all of its input heat energy into work. Some of the input heat energy will always dissipate into the surroundings.

After almost 250 years of development, the heat engine today is more refined. While steam is still the mainstay of large industrial power-generation applications, today most engines run on the combustion of air-gasoline mixtures. Combustion engines power not only automobiles, motorcycles, tractors, ships, and some aircraft, but also lawn mowers and chain saws. Steam engines need boilers, which require external heating—not the most practical arrangement. Combustion engines, also called internal combustion engines, differ in that their fuel combusts directly inside the engine. This allows combustion engines to be smaller and lighter, a necessity in the design of automobiles and other moderate-sized vehicles.

In a typical car engine, a mixture of gasoline and air ignites, explodes, and pushes down on a piston in a cylinder. The translational motion of the piston is converted into rotational motion of a series of machine parts that eventually turn the car's wheels. A cooling system removes excess heat from the engine. A typical automobile has four or more cylinders, which are synchronized so that at least one cylinder performs the power stroke at all times.

We know that no heat engine can be 100% efficient. But what is efficiency? We define the efficiency of a heat engine as the ratio of work output W divided by the heat input Q_{in}.

$$\text{Efficiency} = \frac{\text{Work output}}{\text{Heat input}}$$

In symbols,

$$e = \frac{W}{Q_{in}}$$

Table 7.1 shows the real efficiencies of some selected heat engines.

Table 7.1 Real efficiencies of selected heat engines

Car gasoline engine	10-15%
Steam locomotive	10%
Diesel truck	15%
Solar power plant	30%
Nuclear power plant	35%
Coal power plant	40%

Internal combustion engines are not very efficient because they take several steps to convert heat into work. They have many moving parts and linkages, which are subject to friction and heat loss. In 1824, French engineer Sadi Carnot (1796–1832) showed that the efficiency of a heat engine is limited by the temperature of both the heat that goes in and the heat that comes out. The maximum efficiency possible for any heat engine is expressed in terms of the absolute Kelvin temperatures of the hot-temperature reservoir T_h and the cold-temperature reservoir T_c:

$$\text{Maximum efficiency} = \frac{\text{Temperature of hot reservoir} - \text{Temperature of cold reservoir}}{\text{Temperature of hot reservoir}}$$

In symbols,

$$e_{max} = \frac{T_h - T_c}{T_h}$$

We obtain a greater efficiency if the temperature difference between the hot, or input, region and the cold, or exhaust, region is larger. In another words, we want the input temperature as high as possible and the exhaust temperature as low as possible. However, the temperature of our environment limits the low temperature, and the high temperature cannot be much higher than about 550°C, or the materials used in the engine construction lose their strength and can crack or become brittle. The actual efficiency of a car is not much higher than 10%. **Example 7.3** examines the efficiency of an internal combustion engine in a car.

Example 7.3
Efficiency of an automobile engine
A car's heat engine absorbs 70,000 J of thermal energy and performs 9000 J of work each second. Find (a) the efficiency of the engine; (b) the thermal energy expelled each second; (c) the maximum possible efficiency if the engine is taking in heat at 538°C and dumping heat at 20°C.

Solution:

a. We apply the efficiency relation for a heat engine:

$$\text{Efficiency} = \frac{\text{Work output}}{\text{Heat input}}$$

Therefore,

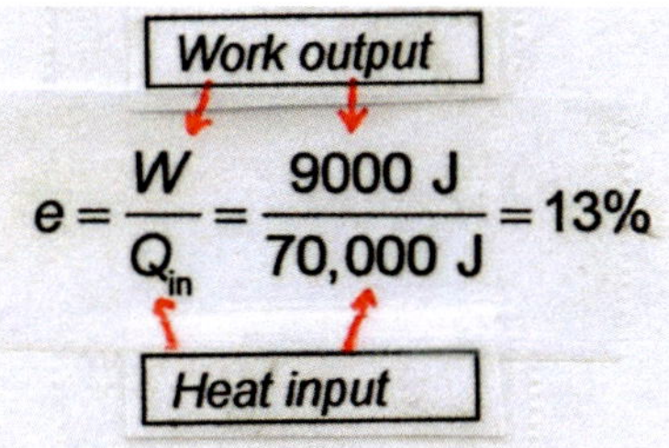

b. Heat input = 70,000 J and Work done = 9000 J. Therefore,

Heat wasted = Heat input – Work done = 70,000 J – 9,000 J = 61,000 J.

The thermal energy expelled is Q_{out} = 61,000 J.

c. First we need to determine the Kelvin temperatures of the engine's heat input and waste output. Recall from Chapter 8 that $T_K = T_C + 273$. Therefore,

T_h = 538°C + 273 = 811 K and T_c = 20°C + 273 = 293 K.

The maximum possible efficiency is

$$\text{Maximum efficiency} = \frac{\text{Temperature of hot reservoir} - \text{Temperature of cold reservoir}}{\text{Temperature of hot reservoir}}$$

$$e_{\text{max}} = \frac{T_{\text{h}} - T_{\text{c}}}{T_{\text{h}}} = \frac{811\text{ K} - 293\text{ K}}{811\text{ K}} = \frac{518\text{ K}}{811\text{ K}} = 64\%$$

The car does not have a very efficient heat engine.

Refrigerators and heat pumps

What would happen if we reversed the natural flow of heat so that it traveled from a cold region to a hot region? We could do this by providing work to extract the heat from a cold reservoir and place it into a hot reservoir. This principle is used in **refrigerators** and air conditioners. In essence, we have a heat engine that runs backward. We have to do work to remove the heat from inside the refrigerator, where the temperature is lower, and place it outside, where the temperature is higher. An air conditioner works similarly: heat is forced from a colder room to the outside of the house, which is hot. See **Figure 7.26**.

Energy distribution in a refrigerator Figure 7.26
It takes work to move heat from the cold reservoir to the hot reservoir, as in a refrigerator or air conditioner.

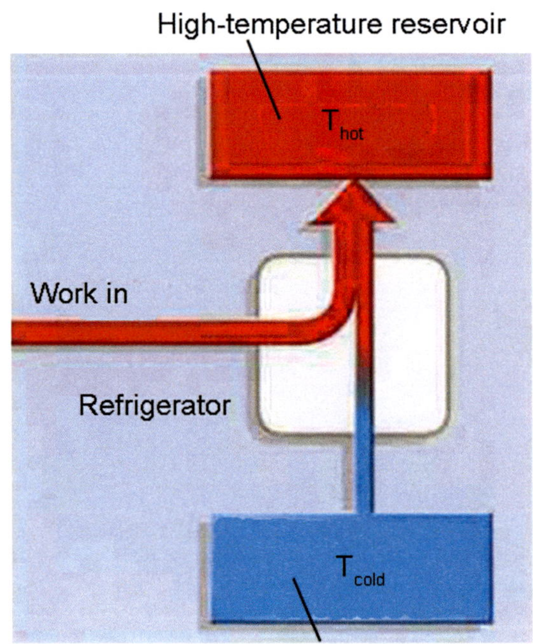

refrigerator A heat engine that runs backward. Work is done to remove heat from inside the refrigerator and place the heat outside the refrigerator.

There is one other difference between heat engines and refrigerators. Heat engines use hot gases as the working medium, while refrigerators and air conditioners use a liquid called a coolant. The coolant has a low boiling point, so it can easily exist as a gas. It must also condense easily when the pressure on it increases.

In the back of a refrigerator, the coolant—in the form of a hot, dense gas—enters a condenser, which is a maze of thin black tubes that draw heat away from the gas and radiate it into the room while the coolant liquefies. The liquid then returns to the evaporator, where it evaporates and draws heat from inside the refrigerator. The coolant now goes to the compressor, which pressurizes the gas so it becomes hot again, and the cycle repeats. In a typical refrigerator, the compressor is doing work on the gas and the source of the energy is electricity. See **Figure 7.27**.

Operation of a refrigerator Figure 7.27
The compressor squeezes the coolant into a hot, dense gas and delivers it to the condenser. In a condenser the coolant gives up heat to the room air and condenses into liquid. The liquid then evaporates inside the refrigerator, extracting heat from the food.

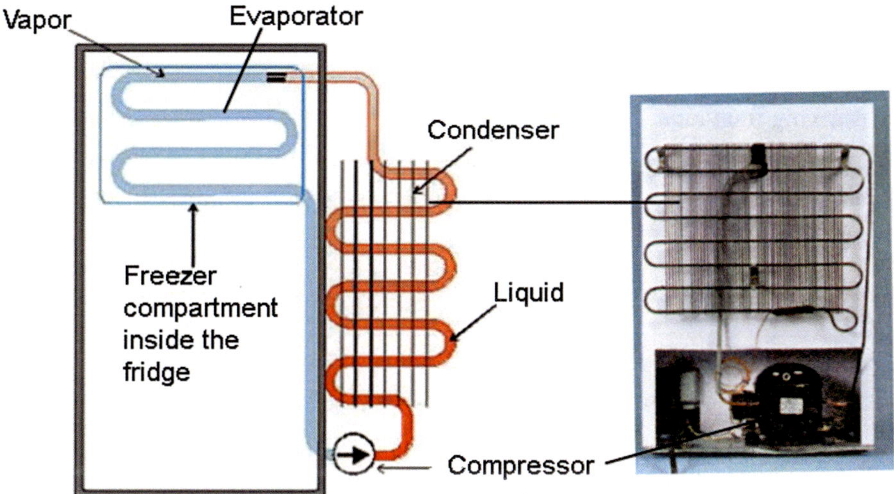

The energy distribution in a refrigerator is as follows:

Low-temperature heat absorbed + work done on the system = high-temperature heat exhaust or

$$Q_{cold} + W_{in} = Q_{ex}$$

This energy distribution means that it is impossible to build a refrigerator that can transfer heat from a lower-temperature region to a higher-temperature region without doing mechanical work. **Example 7.4** gives an example.

Example 7.4
Energy distribution in a refrigerator
How much work does a refrigerator require if it takes in 800 J from the cold region (inside the refrigerator) and exhausts 1400 J to the hot region (air in the kitchen)?

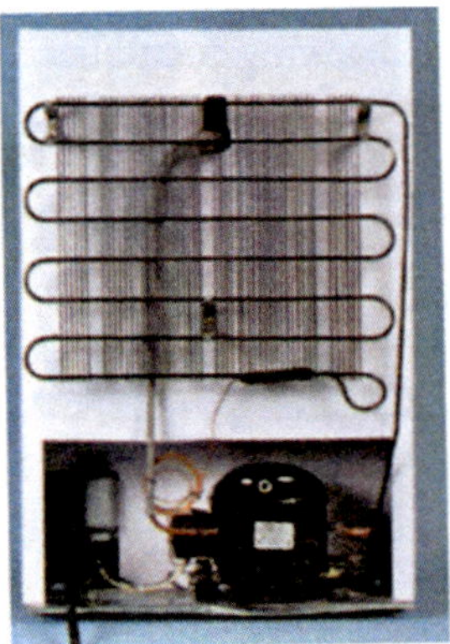

Solution:

We apply the formula for distribution of energy in a refrigerator:

Low-temperature heat + work done on the system = high-temperature heat exhaust
$$800\text{ J} + W = 1400\text{ J}.$$
$$\text{Work done} = 1400\text{ J} - 800\text{ J} = 600\text{ J}.$$

Many homes in moderate climates use *heat pumps,* which work as reversible heat engines. As a cooling device, a heat pump can take heat from low-temperature regions and place it in high-temperature regions. A heat pump can work as an air conditioner during the summer; it extracts heat from the inside of a house and places it outside. It can also function as a heater during the winter, extracting heat from the outside and placing it inside the house. However, heat pumps do not work well when it is very hot or very cold outside.

A popular recent trend is a geothermal heat pump, which is based on the fact that a few feet below Earth's surface the ground remains at a relatively constant temperature, about 10–12 °C (50–54 °F). We can use the ground as a source of heat, taking advantage of its seasonally moderate temperatures.

In contrast, an air-source heat pump draws heat from the air (colder outside air) and thus requires more energy. If the heat pump can exchange heat with the 54-degree ground, it has a much easier time moving that heat in all four seasons because it doesn't have so much temperature difference to overcome in either direction. This results in greater energy efficiency. A ground-source heat pump extracts ground heat in the winter (for heating) and transfers heat back into the ground in the summer (for cooling). The core of the heat pump is a loop of refrigerant pumped through a vapor-compression refrigeration cycle that moves heat similarly to what we saw in **Figure 7.27**.

CONCEPT CHECK

1. **What** is meant by 'directionality' of the second law of thermodynamics?
2. **How** could you make a heat engine run more efficiently?

7.6 Disorder and entropy

LEARNING OBJECTIVE

1. **Explain** dilution of energy.
2. **Describe** entropy.

The second law of thermodynamics states that naturally occurring processes involving heat are directional. That means that they are unlikely to reverse spontaneously. For instance, if you push a book so that it slides across a table, the book will eventually come to a stop because of the force of friction. As the book slides over the surface, its kinetic energy is converted to heat energy. This heat energy cannot be converted back into kinetic energy again: in other words, there is a degradation of energy into heat.

Energy quality and disorder

Thermal energy is less useful than other types of energy. Whenever you transform useful energy such as mechanical energy, you reduce the energy's quality, meaning that it is then much harder to recover that energy and use it again. Once your system creates heat, that system cannot return to its original state spontaneously. For instance, when you ski down a hill, your skis are pressing into the snow, generating heat. Once you get to the bottom, you are not likely to ski spontaneously back up. Instead, you go to the lift that does work on you so that you can go back up the hill. Another version of the second law of thermodynamics could be stated as: *In all natural processes, higher-quality energy has a tendency to transform into lower-quality energy.*

As energy tends to disperse, a higher degree of disorder develops as well. The spontaneous flow of heat from hot to cold regions tends to produce larger disorganization on atomic and molecular levels. To see this, imagine two gases separated from each other in a box with a dividing wall. One gas is hot and the other is cold. Now remove the dividing wall so the two gases come together. The high-temperature molecules have higher kinetic energy. In a collision between the fast "hot" molecules and slow "cold" molecules, the fast molecules lose kinetic energy and the slow molecules gain kinetic energy. Once their kinetic energies and temperatures equalize, the heat flow from hot gas to cold gas stops. On a microscopic scale, the molecules are less organized than before, when each side of the box had molecules with specific average kinetic energy. This microscopic view also shows how on a larger scale, when any two objects are in thermal contact, the hot object cools and the cold object warms over time.

Entropy

We use the term **entropy** to describe the natural dispersing or degrading of energy. We measure entropy by the amount of disorder in a system. What do we mean exactly by order or disorder? A new deck of cards has a high degree of organization or arrangement because the cards are ordered by suit (clubs, spades, hearts, and diamonds) and by number. This is a highly ordered system. Shuffling randomizes the order of the cards, increasing their disorganization. Shuffling the cards back to their original order is highly unlikely. The deck is much more likely to evolve toward a more disorganized state simply because there are so many more ways to disorganize it than there are ways to organize it. In fact, a deck of cards has 8×10^{67} possible arrangements, and only one of them puts them in a perfect order. This is the underlying statistical reason for the increase of disorder—not just of the cards, but of systems in general. See **Figure 7.28**.

entropy Measure of disorder of a system.

Order and disorder in a deck of cards Figure 7.28

A. A deck of 52 cards has 8×10^{67} possible arrangements. Only one of these arrangements puts the cards in a perfect order as shown below.

B. You can shuffle the cards and they will become disordered. If you deal five cards, the chances that you will obtain a winning (ordered) hand are pretty slim. You are a lot more likely to get a "disordered" hand.

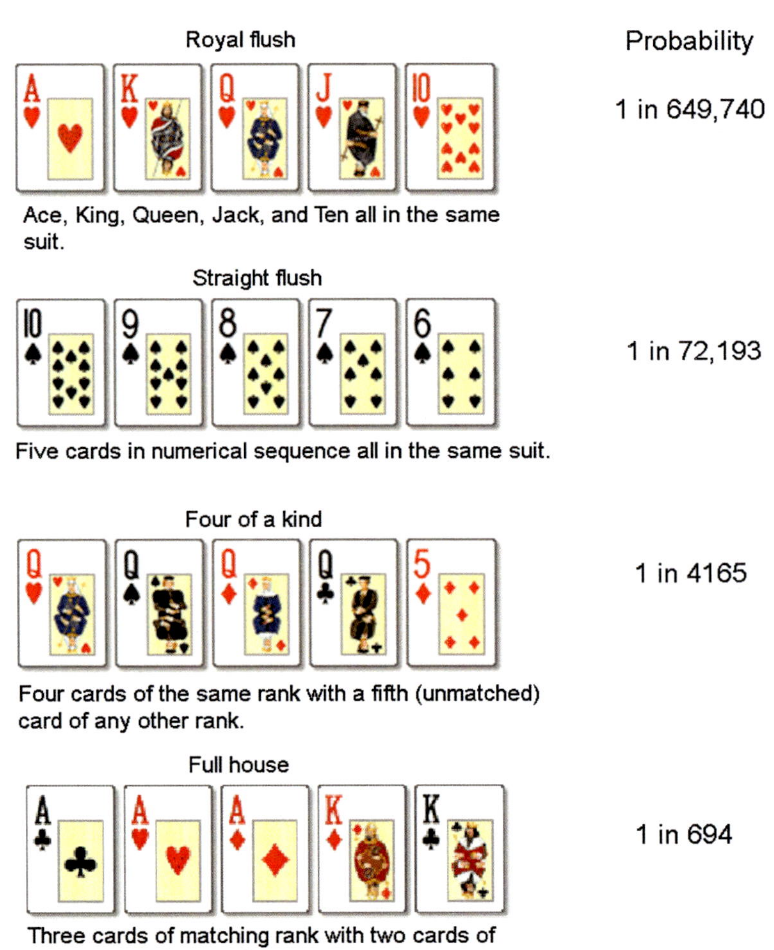

We can describe the second law of thermodynamics in terms of entropy by saying that
The entropy of the universe tends to increase in all natural processes.

For instance, a heat engine takes in heat and puts out energy into two different channels: 1) highly useful and ordered energy in the form of work, and 2) not so useful and disordered energy in the form of waste heat. The expelled heat is at a lower temperature and is difficult to use again. As heat flows from hot to cold, energy dilutes, and disorganization, or an increase of entropy, occurs. Equivalently, an increase in entropy implies a flow of heat from hot to cold. Both are natural processes and go hand in hand. The dilution of energy is as universal as an increase in entropy.

The entropy statement of the second law of thermodynamics is a statement of what is most probable, rather than of what must be. A system naturally tends to move from a state of order to a state of disorder. Unlike energy, entropy is not a conserved quantity. The total entropy in the universe constantly increases as time progresses. Entropy is sometimes called "time's arrow" because time constantly increases as well. This natural tendency of entropy to increase can only be countered by introducing energy in the form of work to straighten things up again.

For instance, if you open a bottle of perfume, the perfume molecules start to escape into your surroundings. This is an irreversible process. As time progresses, it is extremely unlikely that those same molecules would return back to the bottle into a more organized state of higher energy quality. **Figure 7.29** shows what happens to the entropy of a dense gas that is allowed to expand into a vacuum.

Entropy of a system Figure 7.29

A. If you allow gas to expand into a vacuum, the gas becomes more disordered and its entropy increases.

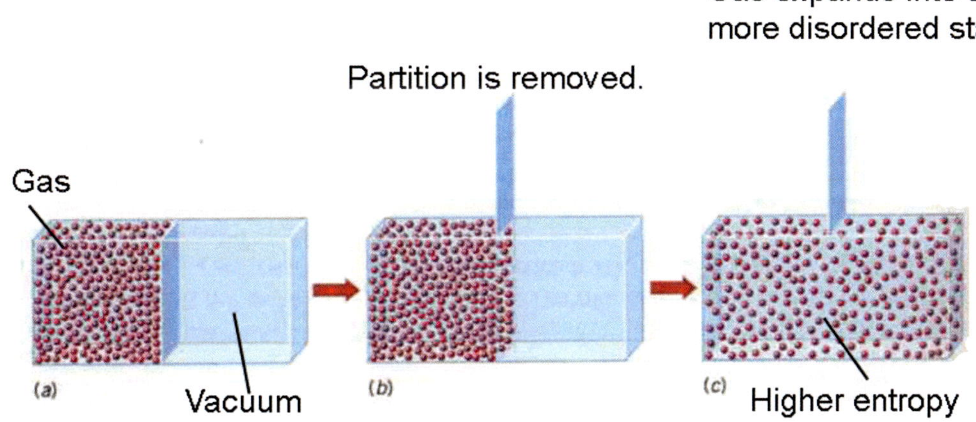

B. If a gas mixture has two different molecules, it is highly unlikely that the molecules will order themselves by kind. It is far more likely that they will mix into a more disordered, higher-entropy state.
C. Similarly, throwing bricks over your shoulder is not likely to create a nice, ordered wall. The bricks have a tendency to reach a more disordered, higher-entropy state.

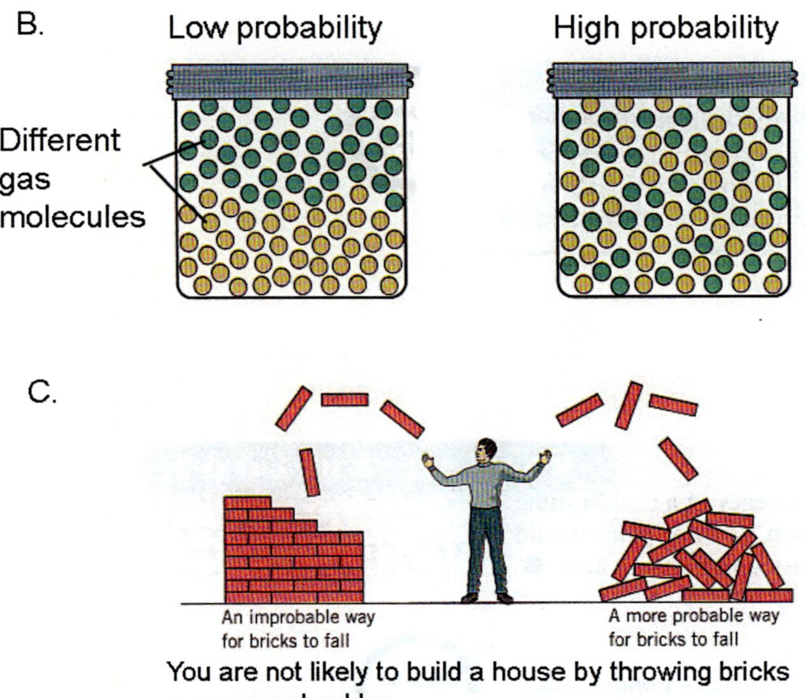

Just as the molecules in **Figure 7.29A** are not likely to all return to the left side of the box, so water in a waterfall is extremely unlikely to run back up by converting its thermal energy into kinetic and gravitational energies, which are higher-quality energy states. If it were not for the second law of thermodynamics, the waterfall and everything else could run backward.

Humans, other animals, and plants are more concentrated and organized systems than the simple ingredients they come from. Is that consistent with the second law of thermodynamics? Two percent of solar energy converts to chemical energy in plant tissues, and the remaining 98% of the energy reradiates back into space. The solar energy flows from hot to cold, and this has a large corresponding entropy increase in the Earth-Sun system. Only 2% of that energy organizes into a low-entropy state, or life's complex, ordered organization. In other words, solar radiation organizes the leaf, but the entropy of the Earth-Sun system still increases.

As heat continues to flow from hot to cold and as disorder continuously increases within our universe, a question arises: What happens when all of this energy flow stops? Nothing much else would happen. In the end, all objects within the universe will have the same temperature. With no temperature differences left, no work can be done because any heat engine needs heat reservoirs at two different temperatures. This was an idea originally proposed by William Thompson, Lord Kelvin, in 1851. He called it the heat death of the universe. His idea suggests that the universe is gradually losing its ability to do work and the total store of usable energy is constantly diminishing.

CONCEPT CHECK

1. **Why** do we lose quality of energy when we drive a car?
2. **How** could you apply entropy to describe the state of your room?

Summary

7.1 Specific Heat and Thermal Expansion

- Energy that transfers from a hotter object to a colder object is called **heat**.
- **Specific heat** is the amount of thermal energy needed to heat up a given amount of a substance by one degree.
- We define the amount of heat need to warm or cool matter as Heat Q = Specific heat $c \times$ mass $m \times$ change in temperature ΔT or $Q = mc\Delta T$
- Objects change their size when their temperature changes. This property is called **thermal expansion** because most objects increase in size as their temperature increases.
- Water expands during cooling from 4°C to 0°C.

7.2 Heat Transfer

- Objects can transfer thermal energy to and from the environment in three different ways: **conduction**, **convection**, and **radiation**.
- Conduction is the transfer of heat energy by the collisions of atoms and electrons within a substance, such as a solid.
- Convection is the transfer of heat energy by moving liquid or gas.
- Radiation is the transfer of heat energy by electromagnetic waves.

7.3 Changes of State

- Most substances can exist in three different forms or *states of matter*: solid, liquid, or gas.
- **Latent heat of melting** is the amount of energy required to change a unit of mass of a substance from a solid to a liquid. **Latent heat of vaporization is the** amount of heat required to change a unit of mass of a substance from a liquid to a gas.
- Water boils at a higher temperature when under higher pressure than normal and at lower temperature when under lower pressure than normal.

7.4 Heat, Work, and Internal Energy

- **Thermodynamics** is the study of heat and its transformations to different forms of energy.
- James Prescott Joule showed that 4.186 J of work produces the same temperature increase as 1 calorie of heat.
- The first law of thermodynamics **states that the increase in internal energy of a system is equal to the heat added plus the work done on the system**.
- Fast expansion or compression of gas where no heat enters or leaves a system is called an **adiabatic process**.

7.5 Second Law of Thermodynamics and Heat Engines

- **The second law of thermodynamics** states that it is impossible for heat to flow spontaneously from a colder body to a hotter body.
- A **heat engine** is a device that takes in heat and converts part of it into useful work. Heat engines use some heat-transfer medium, such as steam or an air-gasoline mixture.
- We can summarize the energy distribution of a heat engine as Heat input = work output + waste heat output
- The second law of thermodynamics when applied to heat engines: It is impossible to have a heat engine that converts all of its input heat energy into work. Some of the input heat energy will always dissipate into the surroundings.
- We define the efficiency of a heat engine as the ratio of work output W divided by the heat input Q_{in}.

$$\text{Efficiency} = \frac{\text{Work output}}{\text{Heat input}}$$

$$e = \frac{W}{Q_{\text{in}}}$$

- The maximum efficiency possible for any heat engine is expressed in terms of the absolute Kelvin temperatures of the hot-temperature reservoir T_h and the cold-temperature reservoir T_c:

$$\text{Maximum efficiency} = \frac{\text{Temperature of hot reservoir} - \text{Temperature of cold reservoir}}{\text{Temperature of hot reservoir}}$$

In symbols,

$$e_{max} = \frac{T_h - T_c}{T_h}$$

- A **refrigerator** is a heat engine that runs backward. Work is done to remove heat from the inside of the refrigerator and place the heat outside of the refrigerator.

7.6 Disorder and Entropy

- The second law of thermodynamics could also be stated as: In all natural processes, higher-quality energy has a tendency to transform into lower-quality energy.
- The spontaneous flow of heat from hot to cold regions tends to produce larger disorder on atomic and molecular levels.
- We use the term **entropy** to describe the natural dispersing or degrading of energy. Entropy is a measure of the disorder of a system.
- The entropy of the universe tends to increase in all natural processes.

Keywords

Heat
Internal energy
Specific heat
Conduction
Convection
Radiation
Solar constant
Latent heat of melting
Latent heat of vaporization
Evaporation
Condensation
Thermodynamics
First law of thermodynamics
Adiabatic process
Second law of thermodynamics
Heat engine
Refrigerator
Entropy

Critical Thinking Questions

1. Water goes into a more ordered state when it freezes in the freezer compartment of a refrigerator. Is this an exception to the entropy principle? Explain.
2. What are the two ways to increase the temperature of an object such as a pot of water?
3. If you shake a bottle of water, what happens to its internal energy?
4. Is it possible to build a heat engine that does not put out any heat exhaust into the environment?
5. What happens to the efficiency of an ideal engine when its input temperature is decreased while the exhaust temperature is kept constant?
6. What does the Second Law of Thermodynamics tell us?
7. How could you increase the efficiency of a heat engine?

8. Describe the difference between a heat engine and a refrigerator.
9. How does the coolant work within a refrigerator?
10. How would you explain an organized organism, such as a plant or a human being, in the context of the law of increasing disorder, or entropy?
11. If a heat engine operated entirely without friction, would it then be 100% efficient? Explain.
12. What is the purpose of a turbine in a power plant?
13. What do you think thermal pollution means?
14. What is the source of inefficiency in a cylinder-based heat engine that converts heat to translational motion?
15. How could entropy be linked to time?
16. When you stick a metal spoon into boiling water, it becomes hot. What type of heat transfer has occurred?
17. What is the optimal position of an immersion heater in a pot? Should you place it closer to the top or the bottom?
18. Why do cats and dogs tend to curl up when they are cold?
19. The rate of heat flow from an object depends on its surface area. What would be the best shape for a house in order to minimize its heat loss in winter?
20. The thermal conductivity of iron is more than five times larger than that of stainless steel. Why do you need to use potholders for pots with iron handles, but not for pots with stainless steel handles?
21. How can moisture contribute to body heat loss?
22. Once springtime comes to the mountains, why is it that all the ice and snow do not melt suddenly, since the temperatures are well above freezing?
23. What materials radiate energy easily?
24. What color clothing would you put on if you wanted to keep warm?
25. What could slow down the greenhouse effect?
26. Is steam more dangerous at 100°C than water at 100°C?
27. Why does an aluminum bench cool down faster than a wooden one?
28. How do thermal conductivity and specific heat enhance heat transfer?
29. Why does a drop of water begin to jump after landing on a red-hot pan?

Exercises

1. You perform 100 calories of work stirring 10 g of water inside an insulated container. The water temperature rises by 9°C. What specific heat of water did you measure?
2. How much heat is released when 100 g of steam at 100°C condenses to water at 100°C?
3. How much heat would it take to melt a 1.5-kg block of ice?
4. How much heat is needed to raise the temperature of 5 kg of copper from 22°C to 962°C (specific heat of copper = 386 J /kg °C)?
5. The Lower Falls in Yellowstone National Park are 94 m high. If all of the water's gravitational potential energy were converted to heat, what would be the increase in its temperature?
6. You mix 1 kg of water at 100°C with 2 kg of water at 50°C in a completely insulated container. What is the final equilibrium temperature of the water?
7. How much heat is required to raise 1 kg of water from 0°C to 100°C?
8. An unknown amount of water at 100°C is mixed with 150 g of water at 50°C in a completely insulated container. If the final temperature of the water is 60°C, what was the mass of the 100°C water?
9. How much heat is required to change 1 kg of lead by 25°C (specific heat of lead = 128 J /kg °C)?
10. A 0.006-kg lead bullet is traveling at a speed of 240 m/s when it embeds in a wooden post. Assume that half of the resultant heat energy generated remains with the embedded bullet. What is its increase in temperature (specific heat of lead = 128 J /kg °C)

11. A heat engine does 350 J of work while exhausting 1000 J. What is its efficiency?
12. A heat engine exhausts 300 J of heat each second. If the engine operates at an efficiency of 30%, at what rate is it able to do work?
13. A steam turbine has a maximum input temperature of 800 K and an exhaust temperature of 300 K. What is its maximum possible efficiency?
14. An engine with a power output of 150 kW operates between temperatures of 500°C and 25°C. If it requires energy at a rate of 500,000 J/s to operate, what are its ideal and actual efficiencies?
15. A person consumes food at 37°C and exhausts waste heat at 20°C. What is the maximum possible efficiency of a heat engine operating between those two temperatures?
16. A swimmer does 550,000 J of work and gives off 340,000 J of heat during a workout. How much internal energy did the body provide?
17. A heat engine has an efficiency of 30%. How much energy must be extracted to do 800 J of work?
18. What is the exhaust temperature of an ideal heat engine that has an efficiency of 40% and an input temperature of 450°C?
19. A refrigerator requires the input of 2,000,000 J of work and gives off 8,000,000 J of heat to the outside. How much heat did it remove from the inside?
20. How much work does a refrigerator that takes in 8,000,000 J from the cold region and discharges 14,000,000 J to the hot region require?
21. A 300-W heat pump delivers heat to your house at a rate of 2000 J per second. How many joules of energy are extracted from the cold air outside each second? Hint: It takes 300 J of electrical work each second to force heat from a cold region outside your house to a warm region inside.